CONTAMINANTS IN AGRICULTURE: EXPLORING THE SOURCES, IMPACTS AND MANAGEMENT

CONTAMINANTS IN AGRICULTURE: EXPLORING THE SOURCES, IMPACTS AND MANAGEMENT

Dr. Hazem Shawky Fouda

Delve
PUBLISHING

www.delvepublishing.com

Contaminants in Agriculture: Exploring the Sources, Impacts and Management

Dr. Hazem Shawky Fouda

Delve Publishing
224 Shoreacres Road
Burlington, ON L7L 2H2
Canada
www.delvepublishing.com
Email: orders@arclereducation.com

ISBN: 978-1-77469-799-3 (Hardcover)

ABOUT THE AUTHOR

Dr. Hazem Shawky Fouda has a PhD. In Agriculture Sciences, obtained his PhD. from the Faculty of Agriculture, Alexandria University, 2008, MSc. In Agriculture Sciences from the Faculty of Agriculture, Alexandria University in 2004, Post-Grade Diploma in Cotton, 2001, BSc. in Agriculture Sciences, from the Faculty of Agriculture, Alexandria University, 1997, worked in Cotton Arbitration & Testing General Organization (CATGO) from 1999 till 2018 was working in the International Cotton Training Center (ICTC) – Cotton Arbitration & Testing General Organization (CATGO) from 2000 till 2015, as a Lecturer & Classer's Trainer for Egyptian and foreigner classers, technicians, ginners, spinners & traders in all cotton aspects. Besides that he was an editor and active member in the Research & Translation Committee, participating in issuing weekly, monthly and annually issues about the international & local cotton market including price trends and direction, recent developments & researches concerning cotton production, protection, harvesting, ginning, fiber testing, spinning & weaving since its foundation in 2000 till 2014 and from 2015 till 2018 he worked as an inspector, since 2018 till present works as a consultant.

TABLE OF CONTENTS

List of Figures .... *xi*

Abstract .... *xiii*

Acknoledgement .... *xv*

Preface .... *xvii*

Chapter 1 The Largest Source of Agricultural Pollution 1

1.1 Introduction 2

1.2 Point Sources 4

1.3 Non Point Sources 15

Chapter 2 Contaminants and Their Effect on Agricultural Crops 21

2.1 Introduction 22

2.2 Pesticides 23

2.3 Mineral Fertilizers 26

2.4 Organic Fertilizers 30

2.5 Wastewater for Irrigation 33

2.6 Agricultural Plastic Waste 34

2.7 Rural Community Waste 37

2.8 Urban and Peri-Urban Agriculture 37

2.9 Accidental Leakages 38

2.10 Electronic Waste 39

Chapter 3 Impacts of Agricultural Contaminants on Both People and Animals 43

3.1 Introduction 44

3.2 Farmers and their Contribution to the Effects of Pesticides on Human Health 45

3.3 Pesticides Associated with Health Risk Factors 46

3.4 Biomagnification and Bioaccumulation of Contaminants in Aquatic Animals 54

Chapter 4 **Crop Contaminant Types, Detection, Monitoring** **67**

4.1 Contamination By Heavy Metals 68

4.2 Food Contamination in Agricultural Soil 72

4.3 Application of Manures and Bio-Solids 74

4.4 Detection 75

4.5 Monitoring 76

Chapter 5 **How Industrial Agriculture Affects Our Water** **87**

5.1 Introduction 88

5.2 Agriculture as a Source Of Water Pollution 88

5.3 Negative Effects of Water Pollution From Agriculture 90

5.4 Rules and Regulations Placed Concerning Water Pollution and Agriculture 93

5.5 Environmental Justice, CAFOS, and Water Pollution 93

5.6 The Scope of the Water Quality Problem 97

5.7 Introducing Agricultural Fertilizers Into Water Supply Systems 101

5.8 The Future 103

Chapter 6 **Agricultural Policy to Contamination** **107**

6.1 Introduction 108

6.2 Establishment of Food Safety Standards and Regulations 109

6.3 Creating Incentives for Food Safety Innovation 111

6.4 Establishment of Liability Standards for Food Safety and Contamination 113

6.5 Environmental Impact of Agricultural Based Contamination and Policies Related 116

6.6 Dangers Facing Policies in the Agricultural Sector Relating to Contamination 117

6.7 Benefits of Certified Food Management Systems 119

6.8 Benefits of Food Safety Training For Employees 120

Chapter 7 **Measures For Protecting Agricultural Soils from Contamination** **127**

7.1 Introduction 128

7.2 Control Technologies for Non-Point Source Contaminants 130

7.3 Position on the Issue of Contaminated Farm Runoff 132

7.4 In Situ Physical/Chemical Treatment Technologies 141

7.5 Bioremediation of Agricultural Soils 142

Chapter 8 Challenges to the Prevention of Agricultural Pollution 149
8.1 Introduction 150
8.2 Challenges to the Prevention of Agricultural Pollution 151

Chapter 9 The Future of Agricultural Contaminant Management 169
9.1 Introduction 170
9.2 Strategies and Implementation 171

Chapter 10 Universal Bodies Pushing For Sustainable Contaminant-Free Agriculture 193
10.1 Introduction 194
10.2 Terms in Universal Sustainable Agriculture Bodies 196
10.3 United Nations Environment Program (UNEP) 203
10.4 Global Alliance on Health and Pollution (GAHP) 207
10.5 European Environment Agency (EEA) 209

References 211

Index 217

LIST OF FIGURES

Figure 1.1: Water Security.

Figure 1.2: Livestock emissions by source. Direct livestock emissions are shown in red.

Figure 1.3: Heavy metal toxicity symptoms in plants.

Figure 1.4: An example of a poorly managed land.

Figure 1.5: Some of the non-point sources.

Figure 2.1: Many millions of hectares are sprayed from the air each year.

Figure 2.2: Fertilizer application on plants.

Figure 2.3: Antimicrobials in soil.

Figure 3.1: Biomagnification of pesticides (DDT) at different trophic levels in a food chain.

Figure 3.2: Polluted sandy beach.

Figure 4.1: Immobilization of heavy metals: an example.

Figure 4.2: Parasites in different foods.

Figure 4.3: Radioactive contamination of food has the most severe effects on human life.

Figure 5.1: Heavy metals in agricultural catchments.

Figure 5.2: Farm pollution explained.

Figure 7.1: Phytoremediation.

Figure 7.2: Bioremediation.

Figure 8.1: Toxic algal blooms.

Figure 8.2: Farmers and fishers along the Mekong River and its tributaries use recession agriculture, a farming system of planting crops on fertile banks as flood waters recede to take advantage of soil moisture.

Figure 8.3: Smart aquaculture.

Figure 9.1: Slash-and-burn agriculture.

Figure 9.2: Farmers' learning process, motivations, and role of education.

Figure 9.3: Crop rotation management in a farm.

Figure 9.4: Pesticide usage.

Figure 10.1: An illustration of sustainable agriculture assessment and evaluation.

ABSTRACT

In recent decades, there has been an increase in the demand for food production. This is attributed to the growth of the human population and the development of new farming techniques. Innovative management solutions are required to make the most out of agricultural inputs while simultaneously reducing their off-site mobility and the impacts they have on the ecosystems.

The primary contributors to the contamination of local soils, sediments, and streams are activities that are associated with horticulture, agriculture, and industrial output. Examples of agricultural pollutants include organic wastes such as manure and decaying plants; runoff from irrigation systems including salts and trace metals; microorganisms; pesticides, herbicides, and insecticides; and chemical herbicides, pesticides, and insecticides.

Utilizing pollution management measures that are both effective and cost-effective is one way to produce more productive agricultural outputs. Applications of fertilizer that are targeted, appropriate, and well-balanced are important if one wants to increase agricultural output while at the same time limiting their influence on the environment. To advance global crop production in a way that is both efficient and kind to the environment, every effort should be made to improve the availability and utilization of secondary and micronutrients, organic fertilizers, and techniques for soil conservation. This should be done without compromising the soil's overall health or its level of productivity. Taking this action is required to increase the crop yields. As a result of this, it is of the utmost importance to find solutions to the challenging issues that crop up on a regular basis in the agricultural industry. As a direct result of the data presented up top, we are thinking of writing a book with the working title "Contaminants in Agriculture: Sources, Impacts, and Management." This book focuses on the many different outcomes that could be brought about by the existing scenario. It is our steadfast opinion that this volume will prove to be a resource that is important for anybody who is interested in agriculture or who is curious about agriculture.

ACKNOLEDGEMENT

First and foremost, I feel always indebted to **ALLAH**, the most beneficent and merciful.

Words stand short when they came to express my gratefulness to my late parents **Shawky Fouda** and **Fatma El-Hussainy**, my uncle **Ahmed Fouda**, my wife **Heba Sadek**, my son **Mohammed Fouda**, my daughter **Raghad Fouda,**

And a special thanks to my sister **Riham Fouda**, relatives and friends for their support, encouragement, care and patience and for their love.

PREFACE

The editor of the book explains why he felt compelled to compile this comprehensive study, which features the most up-to-date information available on the ongoing issues that result from soil and water contamination. This book also features the most up-to-date information available on the ongoing issues that result from soil and water contamination.

The editor worked hard to put together this special volume that is named "Contaminants in Agriculture: Sources, Impacts, and Management,". Heavy metals, excessive use of chemical fertilizers, nutrients, pesticides, herbicides, and insecticides, agricultural wastes, toxic pollutants, and so on all have detrimental effects on agricultural soil and crops; this book focused on those effects and the solutions that can be found for them through various scientific approaches, such as bioremediation.

This book provides an in-depth discussion of the scientific and engineering methodologies, obstacles, and challenges related to bioremediation on a global scale. This book is full of material that will be helpful to a wide variety of people, including agriculturalists, researchers, teachers, agronomists, botanists, and even backyard gardeners.

This book is a must-read for professionals working in the fields of agriculture, agronomy, plant science, plant biology, and biotechnology. This is an added bonus for anyone working in these fields. The book would serve as an excellent resource for readers, scientists, farmers, and others interested in the current state of the art as well as the future directions of research, both in the laboratory and out in the field.

Chapter

1 THE LARGEST SOURCE OF AGRICULTURAL POLLUTION

Contents

1.1 Introduction 2
1.2 Point Sources 4
1.3 Non Point Sources 15

1.1 INTRODUCTION

Agricultural pollution refers to the contamination or degradation of the environment and nearby ecosystems and the harm done to humans and their economic interests, whether by natural or abiotic means. Air pollution, water pollution from a single point of discharge, and nonpoint source pollution, sometimes known as landscape-level pollution, are only a few of the numerous potential sources of pollution. Introducing these chemicals into the ecosystem can have far-reaching consequences, including creating dead zones in vast bodies of water due to the accumulation of agricultural runoff, the extinction of native species, and the poisoning of drinking water (Chen, Y. H. 2017).

The amount of these contaminants and the harm they cause can be drastically altered by the management strategies and the lack of knowledge about them. Many countries worldwide rely on chemical pesticides and fertilizers as agricultural management techniques. Animal housing and management control are other methods of administration. Processes such as improper animal feeding management, overgrazing, plowing, fertilizer use, and the incorrect, excessive, or ineffective application of pesticides all fall under the category of "bad management."

Agriculture-related pollutants considerably affect the quality of all aquatic environments, from lakes and rivers to wetlands and estuaries to groundwater. Sediment, nutrients, viruses, pesticides, metals, and salts are only some contaminants that farming can generate. Pollutant emissions into the environment are greatly affected by the livestock industry. Stream and groundwater contamination can occur from bacteria and pathogens in manure if grazing, dung storage in lagoons, and manure application to fields are not correctly managed. Since changes in land use and animal agriculture practices have disproportionately substantially influenced air pollution, the Special Report of the Intergovernmental Panel on Climate Change (IPCC) on Climate Change and Land has made resolving these concerns a priority.

Any human endeavor, including agriculture, industry, and energy production, can harm the natural environment. Regulatory authorities have identified two primary categories of pollution to safeguard the environment:

- Point-source pollution.
- Nonpoint-source pollution.

In Point-source pollution, it's not hard to trace pollution back to a single point of entry. Its very name suggests a particular geographical origin.

Nonpoint pollution sources on the other hand are more difficult to identify and regulate. It is possible to attribute pollution to several different factors simultaneously.

Point sources are defined by the United States Environmental Protection Agency (EPA) as discrete, localized areas where contaminants are introduced into the ecosystem. Smokestacks, waste pipelines, and ditches are just a few examples. Pollutants from industry and power plants can contaminate the environment in two ways: the air and the water. Particulate matter (small particles), heavy metals, sulfur dioxide, nitrogen dioxide, and carbon monoxide are only some of the toxic pollutants that can be released into the air via smokestacks. When industries like oil refineries, paper mills, and car manufacturers use water in their production processes, they can remove the effluent from waterways. This effluent may contain hazardous chemicals. Wastewater treatment plants in urban areas are also significant pollution hotspots (D'Odorico, P. 2014). The wastewater discharged from a treatment facility risks germs and nutrients being carried downstream in the rivers. Algae blooms can spread uncontrollably when the nitrogen content of water is too high.

Nonpoint-source pollution, like the widespread release of poisons, is the polar opposite of point-source pollution, which results from discrete points of origin. Just imagine what a city street looks like while it's pouring outside. Oil from car engines, rubber bits from worn tires, and dog poop are all washed away by rains washing across the asphalt. After entering a storm drain, the discharge is carried downstream to a river. Most of the pollution from these "nonpoint sources" is caused by runoff. Because of the prevalence of asphalt and tile roofs in metropolitan areas, this is a significant issue. The pollution from a single city block may not seem like much, but when added to the pollution from the hundreds of miles of roadway in a major metropolis, the problem becomes enormous (Chen, Y. H. 2017).

Runoff from logged sections of forest can be used to clean up silt on rural roads. It can also carry acid from abandoned mines and flush chemicals and fertilizer from formerly farmed land. Lakes, rivers, and streams will certainly be contaminated by this waste. Smog and other air pollution are the primary causes of acid rain. When sulfur dioxide, nitrogen oxides, and water molecules come into contact with one another, ozone is produced in the atmosphere. Acid rain is nonpoint-source pollution caused by the long-distance transport of chemicals from several sources, such as industries and power plants.

Enteric fermentation accounts for the vast majority of agriculturally-produced GHG emissions. This activity can be attributed to an annualized carbon dioxide equivalent emission rate of about 179 million metric tons. Through a process known as enteric fermentation, bacteria in the digestive tract break down complex carbohydrates into simpler molecules that the body may absorb and use as fuel. Increased methane emissions can be found in many parts of the world where human agriculture relies heavily on animals whose digestive processes entail enteric fermentation. The different sources of agricultural pollution are discussed below.

1.2 POINT SOURCES

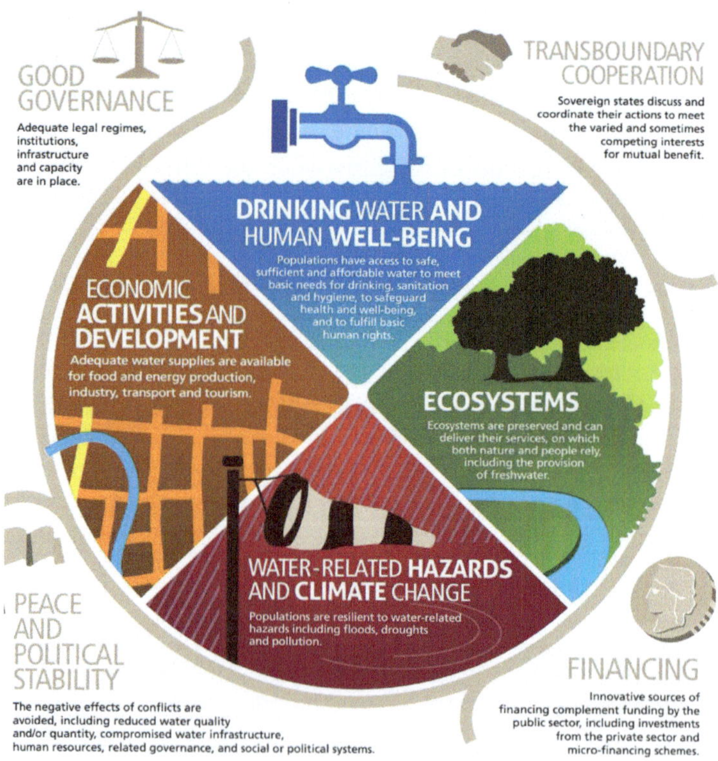

Figure 1.1. Water Security.

Source: Aboelnga HT, Ribbe L, Frechen F-B, Saghir J. Urban Water Security: Definition and Assessment Framework. Resources. 2019; 8(4):178. https://doi.org/10.3390/resources8040178

Contaminated water: One contributor to the agricultural crisis is the contamination of the water used for irrigation. Most of the water that humans consume comes from natural aquifers, man-made canals, and rain. Even though there is a plentiful supply of drinkable water, there is a possibility that some of it contain organic compounds and heavy metals. This has come about as a direct consequence of local industrial and agricultural trash being deposited into water sources in the surrounding area.

Consequently, the crops are subjected to water that contains minute quantities of heavy metals such as mercury, arsenic, lead, and cadmium. When animals are in danger and crop failure is caused by dirty water, it becomes increasingly difficult to resist the harmful effects of agricultural pollution.

The vast majority of domestic sewage comprises pathogens, also known as bacteria, responsible for the disease and organic substances capable of rotting. (Chen, Y. H. 2017). The high probability that contaminated water collected in cities and towns is infected with infectious microorganisms, which are released into the environment through feces, makes it an immediate risk to the general population's health. Organic debris capable of decomposing in water poses a significant threat to water quality. The amount of dissolved oxygen in the water is decreased due to the natural breakdown of organic sewage waste carried out by bacteria and other microbes. The vitality of lakes and streams, where fish and other aquatic species rely on adequate oxygen levels for survival, is jeopardized due to this hazard. Even though sewage treatment processes can reduce the number of pathogens and organic compounds in wastewater, they cannot altogether remove these contaminants.

In addition, sewage that has been collected from residential areas is an abundant source of plant nutrients, particularly nitrates and phosphates. If the levels of nitrate and phosphate in the water are high enough, an algal bloom characterized by algae's rapid and dense development may form. Because bacteria require oxygen to consume dead algae as part of the decomposition process, the amount of dissolved oxygen in the water naturally drops due to this process. Anaerobic organisms are bacteria that can survive and even grow in an environment that does not include oxygen. These bacteria are responsible for the breakdown of organic wastes and are found in anaerobic environments. This process results in the production of potentially hazardous gases such as methane and hydrogen sulfide. It is said that a lake is going through eutrophication when it goes from being clear and clean with low dissolved nutrient concentrations and a balanced aquatic community to having abundant algae and nutrients, then having its oxygen levels drop, and finally having its waste levels drop. This is because eutrophication is responsible for losing a lake's clarity, cleanliness, and low dissolved nutrient concentrations. Eutrophication is a natural process that cannot be reversed and tends to worsen with time. This process can hasten the premature aging and death of a body of water when human activities speed it up and when the water itself is contaminated (a phenomenon known as cultural eutrophication) (Merrington, G., Nfa 2002).

The improper disposal of solid waste is a significant contributor to one of the numerous factors that cause water contamination. The terms garbage, junk, technical waste, trash, and debris from building demolition and construction are all included in the definition of solid waste. In less developed countries, the problem may be made worse due to a lack of resources or regulation that prohibits the inappropriate dumping of junk and a lack of infrastructure for the efficient disposal of solid waste. This could make the situation even more challenging to solve. In certain areas, garbage from landfills is disposed of on purpose in nearby bodies of water (Chen, Y. H. 2017).

Trash and other forms of land pollution can make their way into water bodies and compound the situation. This can happen for various reasons, including animals, the wind, or precipitation. The pollution of the ocean by solid waste from inland waterways may occur in enormous proportions at some point in the future. The pollution caused by solid waste is unsightly and may be hazardous to aquatic habitats; additionally, it may have direct consequences for wildlife. When certain forms of solid waste, such as electronic and plastic garbage, deteriorate and release harmful substances into the environment, this process produces toxic or hazardous trash. These wastes can also originate from other waste streams, such as radioactive waste, amongst other places (D'Odorico, P. 2014).

Livestock: Traditionally, farmers would raise as many animals as their land could comfortably accommodate. The natural diets of cattle, sheep, pigs and other animals were supplemented with materials recovered from agricultural operations. Because of this, the animals contributed to an increase in the farm's overall output. In most cases, cattle are bred in cramped, unsanitary conditions, and they are also forced to consume aberrant diets and are transported to slaughterhouses regularly. They do this, contributing to the pollution problem and the diminishing amount of farmland. Methane (CH4), ammonia (NH3), nitrous oxide (N2O), carbon dioxide (CO2), and nitrous oxide (N2O) are the primary components that make up emissions produced by livestock activities. There is evidence to imply that the carbon dioxide produced by domestic animals considerably contributes to the planet's warming.

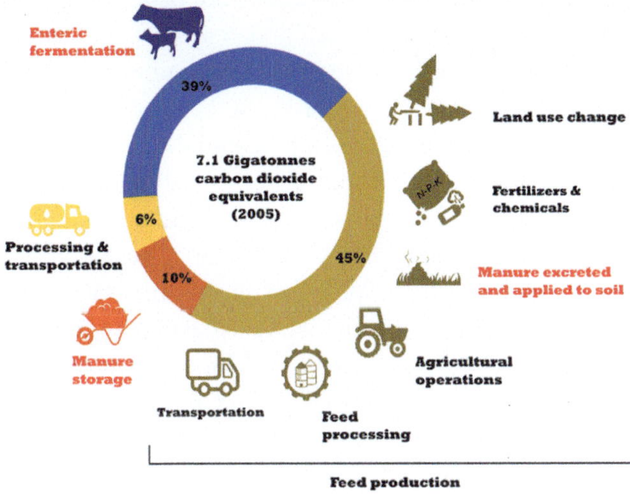

Figure 1.2. Livestock emissions by source. Direct livestock emissions are shown in red.

Source: Livestock and climate change: Impact of livestock on climate and mitigation strategies - Scientific Figure on Animal Frontiers. Available from: Giampiero Grossi, Pietro Goglio, Andrea Vitali, Adrian G Williams, Livestock and climate change: impact of livestock on climate and mitigation strategies, Animal Frontiers, Volume 9, Issue 1, January 2019, Pages 69 76, https://doi.org/10.1093/af/vfy034

Carbon dioxide emissions from cattle production are expected to have surpassed the 565-gigatonne level by the year 2030, according to projections made by scientific researchers. The production of animals is responsible for 68 percent of all enterogenic emissions of nitrous oxide. This gas has the potential to remain in the atmosphere for up to 150 years, during which time it can warm the earth to a temperature 296 times higher than that caused by carbon dioxide and cause damage to the ozone layer. Animal agriculture is responsible for 64 percent of all ammonia emissions, making it one of the most significant contributors to acid rain and the acidification of ecosystems. In addition, cattle are responsible for 35 and 40 percent of the world's methane emissions. Carbon dioxide has a probability is 23 times greater than that methane, being a contributor to the warming of the planet.

According to data provided by the EPA, animal production was a contributing factor in a 50% increase in methane emissions over the course of the preceding 15 years, with pigs being responsible for a 37% increase in this increase. Second, some types of livestock are better for the environment than others, but growing animals for food may be very space-intensive. Currently, more than two-thirds of the world's agricultural land and one-third of the world's total land area is used for livestock production. The livestock sector has emerged as the principal force behind the clearing of forests and the transformation of pastureland into crops as a direct result of the rising demand for animal products combined with the limited availability of land. According to the International Center for Forestry Research (CIFOR), the quantity of land converted to pasture during this period was sufficient to divide Portugal in half. It has been documented that humans have been known to cut down trees to provide animal food. More than forty percent of the world's cultivated crops are solely used to feed animals. If half of the crops currently used to feed animals were instead utilized to feed people, then all people who are now in need of food would be provided for (Merrington, G., Nfa 2002). The massive cutting down of trees has led to the destruction of habitat. Deforestation is responsible for the extinction of anywhere from 137 to 179 unique species of plants, animals, and insects daily. This is the most catastrophic mass extinction that has taken place in the last 65 million years, according to Ceballos et al. even though cattle ranching and other forms of animal husbandry are frequently held responsible for adverse environmental impacts, arid and semi-arid regions are home to some of the most innovative approaches to food production. This is the case even though there is universal consensus that the cattle business and livestock farming are responsible for or contribute significantly to various environmental problems. Because of increased climate variability, it will be more challenging to forecast when and how much food will need to be produced. As a result, agriculture may not be the best land use, and other options should be considered. Assuming that animals can move about freely, this land utilization would be more productive and make more sense than farming. Rice is one of several staple crops known to have disastrous effects on the settings in which they are grown (Chen, Y. H. 2017).

The production of animal feed is the most significant contributor to the contamination of water sources in agricultural regions. The growing demand for animal products is devastatingly affecting the world's ecosystems and water supplies, particularly in less developed nations. The chief contributors to water contamination include animal waste, hormones and antibiotics, fertilizers and pesticides used in fodder production, and pasture precipitation runoff. Animal waste and the manure left behind by chickens are two of the most common contributors to the contamination of water sources, as stated by the United States Department of Agriculture (USDA).

In addition, the cattle industry is a significant contributor to the waste of resources and a substantial consumer of water. Over 55 percent of the country›s freshwater supply is used for animal husbandry in the United States, whereas only about 5 percent of that supply is used for residential use. According to the findings of a study that compared the effects of various diets on the environment, a vegan diet has the least detrimental impact on the planet when taking into account factors like the number of resources that are required (such as water, fertilizer, and soil), as well as the amount of greenhouse gas emissions that are created by food. A vegan prevents the unnecessary death of one animal each day while conserving 4164 liters of water, 20 kilograms of crops, 2.8 square meters of forest per day, 10 kilograms of carbon dioxide per day, and 2.8 square meters of forest per day.

In addition, the cattle business creates a significant amount of waste. The livestock industry in the United States generates around 116,000 pounds of waste every second. According to the findings of the research conducted by Haines and Staley, the amount of garbage that a farm with 2,500 milking cows generates is comparable to that generated by a metropolis with 411,000 residents. As a result, we must determine whether or not the general population is aware of the environmental damage produced by the cattle industry.

Heavy metals: Using fertilizers, manure, and other organic wastes, including heavy metals, can accumulate contaminants in the soil. This buildup can have negative consequences for human health. Selenium can accumulate as a result of agricultural practices like irrigation.

The consumption of these chemicals is hazardous to one's health and can potentially expedite death among people. When leached into the water, taken up by plants, or washed down the drain, they eventually make their way into streams and groundwater. Heavy metals present in contaminated water or food can severely affect the livestock. Heavy metals can also cause considerable harm to crops or even cause them to fail to yield a harvest. A variety of contributing elements is making the environment worse, including the rapid growth of the economy and the proliferation of various industries, such as manufacturing and agriculture.

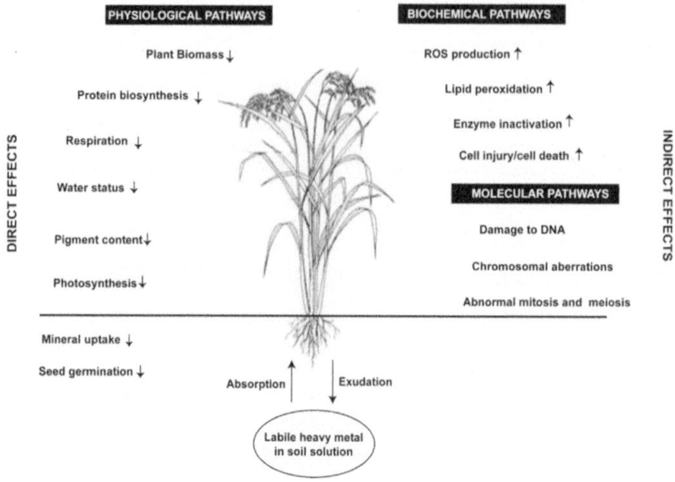

Figure 1.3. Heavy metal toxicity symptoms in plants.

Source: Bhat JA, Shivaraj SM, Singh P, Navadagi DB, Tripathi DK, Dash PK, Solanke AU, Sonah H, Deshmukh R. Role of Silicon in Mitigation of Heavy Metal Stresses in Crop Plants. Plants. 2019; 8(3):71. https://doi.org/10.3390/plants8030071

Both natural and man-made processes can result in the introduction of potentially harmful chemical compounds that are collectively referred to as environmental pollutants. Several environmental actions could jeopardize biotic populations (including humans, plants, and animals) and abiotic communities. Synthetic industries, coal conversion, and waste incineration fall into this category. Pesticides and heavy metals are two of the most prevalent environmental toxicants, and both pose a significant risk to ecosystems' overall health and functionality. Heavy metals are defined as those with a high atomic weight and densities greater than 5 g/cc. The chemical features of heavy metals ought to be valued more highly than their physical characteristics. Think tanks worldwide are starting to pay more attention to environmental toxicity that exceeds the maximum residual limits authorized. (MRL). It is widely known that the hazardous metals cadmium (Cd), lead (Pb), copper (Cu), and zinc (Zn) interact with one another. Mining, industrial production, and even agricultural practices can all be a source of heavy metals. Examples of supplies that originate in the agricultural industry include wastewater, fertilizer, pesticides, and animal excrement. Poisoning caused by heavy metals in the environment is a developing worry that can have severe repercussions, particularly in the agricultural sector. The buildup of metals in the soil and the subsequent uptake of those metals by flora are the primary factors contributing to this problem. The pollution caused by heavy metals requires the development of fresh strategies that may be implemented in various contexts to alleviate the associated issues (Merrington, G., Nfa 2002).

Heavy metals are essential to the healthy functioning of various plant and human organs, but they become hazardous once they reach a particular concentration. Heavy metals are essential to the proper functioning of a variety of plant and human organs. According to the findings of multiple investigations conducted in this sector, mining, agricultural chemicals, industrial activities, and agricultural practices have all been found to be substantial contributors to the heavy metal load in the environment. The research undertaken by Xiao and colleagues shows that both the agricultural and industrial sectors considerably impact the amounts of heavy metals found in agricultural soil and crops. This is especially the case near businesses dealing with cement and electroplating. To put it another way, the topmost layer of the soil is an ideal location for accumulating heavy metals, which the plant's roots and vascular system can subsequently absorb together with water (Chen, Y. H. 2017).

The accumulation of heavy metals is comparable to the accumulation of different components of an ecosystem. The roots of plants are the primary sites where heavy metal ions taken up from the soil interact with one another. They typically carry out this role to minimize the bioavailability of poisons in the soil by binding and then fixing them. Phytoextraction is a method that utilizes plants to extract toxins from polluted soil to restore the environment. Rhizofiltration and phytostabilization are two other methods that can be used to give plants access to heavy metals. Through a process known as phytostabilization, it is possible to limit the mobility of heavy metals in the soil as well as immobilize the metals themselves (a form of phytoremediation to use plant roots to absorb toxic substances). Both humans and plants are susceptible to becoming ill after ingesting these metals in their ingested form.

The buildup of toxins in the food chain and their continued presence in the environment contribute to worsening the cumulative effects of toxins on human health. These pollutants have been associated with a variety of acute and chronic disorders in humans, including lung cancer, renal dysfunction, osteoporosis, and heart failure. Tong et al. researched the severity of the dangers that heavy metals pose to the health of urban people in China from 2003–2009. According to a study that investigated the dangers that heavy metals pose to human health, absorption is the most common exposure associated with adverse effects on human health. This was determined based on the findings of the study. The accumulation of heavy metals in human tissues has been linked to an increased risk of several ailments, including migraines, epilepsy, and comas. These metals may also play a role in the genesis of further diseases by acting as pseudo-cofactors. Heavy metals are another factor that might contribute to the death of tissue. There is a correlation between exposure to heavy metals and an increased risk of sickness in children and adults. Ingestion (via food, drink, or breath) or skin contact with surfaces contaminated with pesticides can lead to adverse health effects in humans and other organisms. Pesticides can harm human health, even with minimal exposure. The use of pesticides has also been linked to a variety of additional health issues, including hypersensitivity, cancer, asthma, and abnormalities in hormone levels. They have been linked to infant mortality, low birth weight, and congenital anomalies.

With these environmental issues in mind, our research aims to discover the sources of heavy metals and pesticides, as well as the many different forms that these substances might take. In addition, research is conducted to investigate how chemical and heavy metal contamination affect the physical qualities of agricultural soil. In addition, research has been done on the potentially lethal consequences of various plant species. Notable because it offers a complete comprehension of the processes through which heavy metals and pesticides enter the environment, accumulate in the soil and plants, and finally enter the bodies of humans. Heavy metals and pesticides have been proven to be both synergistic and antagonistic, with reports of their combined toxicity in soil, plants, and humans. This is because heavy metals and pesticides have both been demonstrated to interact with each other.

Organic Contaminants: Using manures and biosolids in agriculture almost always results in soil nitrogen, carbon, and phosphorus alteration. Because of the industrial nature of both medications and cosmetics (PPCPs), there is a possibility that they will become contaminated. Because it has been demonstrated that these compounds can be found in humans and animals, researchers have shown that they may threaten the health of humans, animals, and the environment (Chen, Y. H. 2017).

The existence of organic pollutants adds a major complexity to the efforts made to regulate agricultural pollution. The use of herbicides and insecticides, as well as the decomposition of animal and plant tissue, are all examples of organic pollutants that have a detrimental effect on ecosystems. It is possible that the existence of organic pollutant residues in the food supply, as well as in the land, water, and air, may have negative impacts on both human health and the health of the ecosystem. Because biochar is capable of surface adsorption, partitioning, and sequestration, it can reduce the negative impacts that these organic pollutants have. During the adsorption process, a contact reaction between the charcoal and the organic pollutant results in the chemical bonding of the contaminant to the charcoal. Biochars with a sizable surface area and a consistent pore size distribution are ideal for usage in soils that need to be cleansed of organic contaminants. It has been demonstrated that biochar can increase the number of adsorbed organic pollutants and reduce the number of desorbed pollutants through various mechanisms, including interaction and pore obstruction.

Pollutants and wastes that have an organic composition are made up of carbon and hydrogen far greater than any other elements. Other elements are present but to a much lesser degree. Organic pollutants known as volatile organic compounds (VOCs) can be found in various gaseous or liquid substances, such as solvents, chemical precursors and intermediates, petroleum products, and many more. VOCs are known to be harmful to human health and the environment. Sludges, still bottoms, resins, chemicals, waxes, paper, plastic, and wood are some examples of solid organic compounds that can be found. Because their undesirable properties (such as toxicity, flammability, and so on) are a result of the structure of the compounds rather than the elements that compose them, organic compounds can be removed from contaminated media and destroyed in large amounts through processes that alter their structure, such as oxidation or thermal decomposition (Merrington, G., Nfa 2002). This is in contrast to inorganic compounds, which cannot be removed from contaminated media or destroyed in large amounts through such processes. As a result of the fact that these processes can alter the structure of organic molecules, it is possible in theory. Electrostatic filtering and condensation at low temperatures are two more methods that can be utilized to remove gas-phase organic molecules from a gaseous medium, provided that the parameters outlined above are satisfied.

Poor land management

Figure 1.4. An example of a poorly managed land.

Source: ISRIC. (2012). Calculating productivity loss due to land degradation. Retrieved from ISRIC World Soil Information: https://www.isric.org/projects/calculating-productivity-loss-due-land-degradation

Poor land management leads to an irreversible decline in soil fertility. Profound land management is crucial for keeping agricultural pollution to a minimum level. Therefore, farmers should be aware of how their actions can impact the environment. Buildings, roads, and parking lots on land all result in impervious surfaces. The ability of the soil to filter runoff is reduced by the presence of impervious surfaces, which contributes to nonpoint source water pollution. As a result, dirtier runoff is produced. The risk of erosion will increase, the ecosystem will change, and the water quality will suffer if the water volume and peak flow are changed. To ensure that more pollutants infiltrate water sources that are used for human consumption and recreation, it is recommended that stormwater outflow be increased. Storm runoff from cities and suburbs contains various harmful elements, including mud, oil from paved surfaces, nutrients from fertilizers, and more. The rate at which aquifers are recharged by groundwater should be adjusted. Point sources, such as municipal and industrial wastewater treatment plants, may release warm water and harmful chemicals in their discharges. Suburbanization and other forms of distributed land development have the potential to exacerbate a variety of environmental issues. Just as an example: The higher concentrations of air contaminants caused by the growing usage of automobiles in urban areas may exacerbate preexisting illnesses like asthma. Urban and suburban "heat islands," domes of warmer air, can occur when trees and bushes are removed, adding to the increased heat absorption by pavement, buildings, and other sources. Heat islands can affect not just the local, regional, and global climate but also the quality of the air we breathe. The quantity and quality of water in watersheds may be affected by the following forms of agricultural land use. The amount of water accessible for different uses can be affected by the crops grown, the cultivation methods employed, and the types of irrigation systems employed. Livestock grazing in riparian areas can change the ecology by removing vegetation from the stream bank and raising sediment and nutrient levels in the water. Vegetation near stream banks is less likely to regrow after being grazed by animals (Merrington, G., Nfa 2002).

Humans are in danger of being exposed to a wide variety of particles and poisons because agriculture may contribute to the deterioration of natural ecosystems and the increase in dust and wind erosion. Invasive species could spread quickly or even be made worse by certain land uses. Overgrazing, land conversion, fertilization, and agricultural pesticides are examples of agricultural land use practices that can hasten the spread of invasive species. They cause harm to humans and animals alike, change the conditions in which fish and other organisms can survive, and are a major cause of the decline in biodiversity. When invasive species are introduced to agricultural settings, they have the potential to reduce water quality and reduce access to water for local fish and animals. Science is beginning to shed light on infectious illnesses' consequences and changing land use trends. One study's findings suggest that changes in land use and other environmental factors may contribute to the spread of vector-borne diseases. The results of another study indicate that the "edge effect" is exacerbated when a forest's habitat is fragmented and separated from urban and agricultural areas. As a result, there will be more disease transmission between different hosts. (Chen, Y. H. 2017). The restoration of polluted land for residential or commercial development and the creation of new wildlife habitats are two examples of the positive outcomes that can occur from changes in land use (as a direct result of operations designed to restore habitat).

1.3 NON POINT SOURCES

Pesticides, pests and fertilizers: Most people think the contamination is caused by using different pesticides and fertilizers. For insecticides and fertilizers on the market to work, they need to be used not only on native species that have been there for a long time but also on exotic species that are moving into the area.

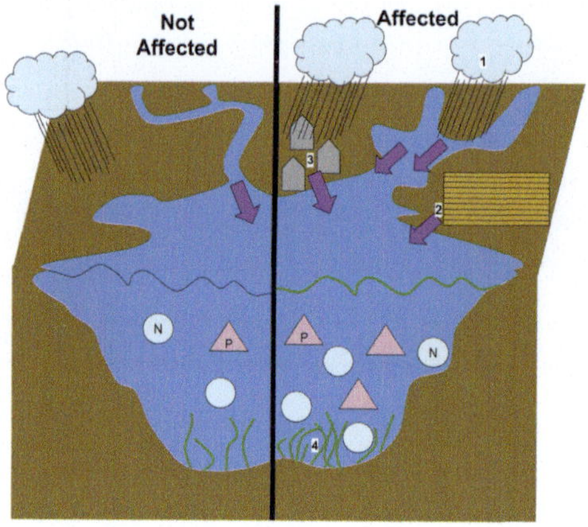

Figure 1.5. Some of the non-point sources: Nonpoint source pollution is caused when precipitation (1) carries pollutants from the ground such as nitrogen (N) and phosphorus (P) pollutants which come from fertilizers used on farm lands (2) or urban areas (3). These nutrients can cause eutrophication (4).

Source: By Houellette - Own work, CC BY-SA 4.0, https://commons.wikimedia.org/w/index.php?curid=59859874 **s**

Pesticides have been an important tool for farmers throughout the history of agriculture. They protect crops and livestock from insect attacks at the cost of overall crop productivity. Even though pesticides are useful, they pose a threat to the natural environment and the health of all living things, and the safety of the food supply. Concerns about the effects of pesticides have made people want to learn more about what happens to chemicals in the environment. Chemicals can move from fields that have been treated to the areas around them through the wind, air, and water. Without a doubt, poor countries need pesticides for farming. Several studies show that health and the environment have become major concerns for these countries. In the last fifty years, pesticides have helped the world make more and better food.

As their use has grown, so have worries that they could hurt things like humans that weren't meant to be hurt. Because pesticides are so common in the natural world, they can hurt all living things, including humans and other animals. Even though pesticides can be used in different ways that could help these organisms, this is still the case. Pesticides can damage people's health, and some studies have even found a link between being exposed to pesticides and getting cancer. Some examples of these types of cancers are solid tumors, non-lymphoma, Hodgkin's, Hodgkin's, and non-Hodgkin's leukemias. In recent years, people have been using dangerous chemicals and pesticides more carelessly, making people worry more about the general population's health. MRLs for food contaminants have only been used by a small number of countries, regions, and international organizations worldwide. Several countries worldwide have set up national food pesticide monitoring systems for various reasons, such as protecting consumer health, improving resource management, and reducing economic losses (D'Odorico, P. 2014).

Two kinds of compounds could be dangerous: toxins and insecticides. Luckily, more and more people are becoming aware of the risk, and companies are working hard to make safer chemicals and more effective ways to use them. Even though there is more information about the risk-benefit analysis, the research has not yet been finished. Pesticides are sprayed in the wrong way almost all of the time. But even when pesticides are used correctly, they may still affect the environment and unwanted animals. Pesticides can have good and bad effects, depending on their use. In this study about pesticides, some of the things discussed are how pesticides are grouped, how they affect people's health, and how they affect the environment.

Pesticides, in particular (herbicides, insecticides, and fungicides), have recently seen widespread use due to the rapid development of technology. This poses a hazard not only to the environment but also to the humans that occupy it. Several of these substances have been implicated in the development of cancer. There is a correlation between being exposed to multiple of these chemicals and having abnormal births. Pesticides are chemicals manufactured and used to control pest populations in various situations, including agriculture. When integrated pest management (IPM) programs use pesticides, they are viewed as useful instruments that are also cost-effective and efficient. Due to bioaccumulation in food chains, pesticides can threaten mammals and other organisms that were not originally meant to be harmed by them. Furthermore, pesticides are ecologically disruptive because they may have unintended impacts on organisms other than those they were meant to target. This is because pesticides are designed to treat specific organisms. Pesticides leave behind environmental residues, which can be discovered in the water, soil, and air. These residues pose a significant threat to the ecosystem because they are carcinogenic and can survive in the environment. The elimination of native plant and animal species in favor of expanding non-native crop production has been an industry standard for a long time. However, this will make the problem of agricultural pollution, which is currently rather significant, even worse. Native people are now faced with new challenges, such as sickness, pests, and weeds, as a direct result of the introduction of new crops.

Because of this, invasive species pose a significant risk to ecosystems because they are responsible for the extinction of native plants and animal life. This is especially true of genetically modified organisms (GMOs), which produce plant and animal species that can wipe out other species within a few short years. As a direct result of the rapid expansion of technological capabilities, humans have been subjected to a wider variety of broad-spectrum chemicals in recent years, leading to greater levels of chemical exposure overall. The category of chemical substances known as "pesticides" is the subject of the noun "pesticide." Throughout agricultural history, pesticides have been an indispensable tool for farmers, serving to guard crops and livestock against insect attacks at the expense of overall crop productivity. Despite the fact that they serve a beneficial purpose, pesticides pose a threat not only to the natural environment but also to the health of all forms of life and the security of the food supply. Research on the environmental fate of chemical agents has been stimulated as a result of concerns about the impacts of pesticides. Chemical agents can move from treated fields to the areas surrounding them by the wind, air, and water. There is no doubt that agricultural pesticides are necessary for developing countries. Despite this, several studies suggest that issues relating to human health and the environment have become significant concerns for these countries. It has been possible for the globe to produce more food of higher quality over the last fifty years thanks in major part to the widespread use of pesticides.

Soil Erosion and sedimentation: Soil erosion and silt buildup are two other things that can cause trouble. Agriculture and grazing are only allowed on the very top layer of the soil, which is composed of several layers. Due to inefficient farming methods, this area loses a lot of soil every year, which causes the nutrients in the soil to be used up. This stuff must have landed in one place at some point, and it could have happened because of wind or water.

Sedimentation is the slow dirt buildup in low-lying areas like ditches, rivers, streams, and farms nearby. Because of this, agricultural pollution stops water, aquatic life, and nutrients from naturally moving from one fertile area to another. Excess nutrients: It is possible for manure and fertilizers, particularly those high in phosphate and nitrogen, to include excessive amounts of chemical nutrition. This can cause agricultural supplies to become contaminated. An excessive amount of nutrients harms the biological integrity of aquatic ecosystems and the inherent quality of the water they inhabit (Merrington, G., Nfa 2002).

During wetter seasons, nutrients have a greater chance of being carried into bodies of water like rivers, lakes, streams, and even the ocean. Due to alterations to the marine and freshwater nutrient cycles, related ecosystems may see shifts in the species makeup of their resident fauna. Anoxia, also known as a deficiency of oxygen in the water, is the most typical consequence of eutrophication. This condition, lethal to fish and other aquatic species, is the most prevalent cause of death. Inadequate management in the agricultural industry can contaminate the surrounding environment with fertilizer runoff. The most prevalent causes of nutrient contamination in agricultural areas include the use of fertilizers that are rich in both nitrogen and phosphorus, as well as the presence of animal feces. When it rains, water and soil containing nitrogen and phosphorus are more likely to wash into nearby waters or drain into groundwater, which degrades the quality of the water.

Compounds based on gaseous nitrogen can be created in potentially substantial quantities by cattle and fertilized soils (Merrington, G., Nfa 2002). include the chemical compounds ammonia and nitrogen oxide. There is a possibility that the aquatic life in surface rivers will suffer as a result of large-scale releases of ammonia. N2O is a greenhouse gas with an extremely high potential to contribute to the planet's warming.

An excessive amount of nutrients can lead to various issues, including a reduction in the amount of oxygen in the water. Algae development has become so dense that it prevents sunlight from reaching the seagrasses and other plants necessary for their growth. After they have passed away, seagrass and algae undergo the natural decomposition process as a part of their life cycle. During the process of decomposition, oxygen from the water is consumed, which leads to a decrease in the amount of oxygen dissolved in the water. This may be fatal for a variety of marine organisms, including fish, crabs, and oysters, among others. Foods that are high in nutrients can be obtained from a wide variety of various places. These dangers could be caused by several different things, such as the interplay of several different water currents or the natural erosion of rocks and soil in the watershed. Because the levels of natural inputs are so much lower compared to human inputs, researchers are extremely interested in the nutrient contributions that coastal residents offer. Farms, urban runoff during storms, and wastewater treatment plants contribute to the number of nutrients discharged into coastal oceans. This is a direct result of the increasing human density along the coast. Any of the manufacturing as mentioned above procedures have the potential to introduce nutritional contamination (D'Odorico, P. 2014).

Chapter

2 CONTAMINANTS AND THEIR EFFECT ON AGRICULTURAL CROPS

Contents

2.1 Introduction 22
2.2 Pesticides 23
2.3 Mineral Fertilizers 26
2.4 Organic Fertilizers 30
2.5 Wastewater for Irrigation 33
2.6 Agricultural Plastic Waste 34
2.7 Rural Community Waste 37
2.8 Urban and Peri-Urban Agriculture 37
2.9 Accidental Leakages 38
2.10 Electronic Waste 39

2.1 INTRODUCTION

Demand for crops and animals has risen over the past several decades as a result of population growth and technological developments in agriculture. To maximize the benefits of agricultural inputs while minimizing their off-site mobility and negative effects on recipient environments, innovative management approaches are necessary.

Industrial, horticultural, and other agricultural activities are the primary sources of pollution in the surrounding soils, sediments, and waterways. Salts and trace metals left over from irrigation, organic and inorganic pesticides, herbicides, and insecticide residues, as well as microorganisms and organic elements such as animal feces and decaying plant matter, are examples of agricultural pollutants.

Effective solutions for the control of agricultural pollutants pave the path to overcoming obstacles to boosting crop yield. To maximize yields without affecting the environment, fertilizer application must be targeted, sufficient, and balanced. Without compromising soil health or productivity, every effort should be made to improve the availability and use of secondary and micronutrients, organic fertilizers, and soil-conservation techniques to develop overall crop production in a way that is both environmentally responsible and effective. Thus, we must address the complex issues that always arise in the agricultural economy.

Only three crops (wheat, maize, and rice) account for about 60% of total dietary energy consumption. In recent years, rising levels of emerging contaminants (EC) in agricultural products have been a significant reason for worry. The contamination of agriculturally viable land and water resources by human activities, technical improvements, and resource extraction reduces soil quality and crop production. Pharmaceutical drugs, insecticides, and mycotoxins are examples of man-made or naturally occurring molecules or substances that are toxic to living organisms. ECs are biologically active substances that, if present in the agricultural system, may have deleterious impacts on plants, soil, and ultimately human health (Fairbairn et al, 2016).

Both biotic and abiotic variables have been shown to influence agricultural plants' capacity to absorb and accumulate ECs. Abiotic impacts include EC chemical properties and environmental disturbances. Vegetables with the greatest potential to absorb and retain ECs are leafy vegetables, followed by root vegetables, cereals and forage crops, and fruit vegetables. Because ECs can disrupt plant physiological and molecular functioning and drastically reduce agricultural production, greater research on crop species is required. The existence of ECs in soil-plant systems poses difficulties for agriculture research. To sustain agricultural output, it is necessary to comprehend how ECs affect crop yields and nutritional quality. This chapter investigates diverse EC types, their interactions with soil-plant systems, and their effects on agricultural productivity.

2.2 PESTICIDES

Pesticides are defined as "any combination or mixture of chemicals made of chemical or biological elements intended to repel, kill, or control any insect or to govern plant development" in the International Code of Conduct on Pesticide Management. Compounds sprayed on crops before or after harvest to prevent spoilage during storage and transport fall under the category of pesticides. In addition, there are chemicals used for fruit sets, fruit thinning, and preventing the early drop of fruit. Global pesticide usage has the potential to harm soil organisms, animals, and non-target plant species in addition to human health. Additionally, pesticides are applied to manage disease vectors in people and animals.

For as long as there has been agriculture, there have been pests and crop illnesses, and humans have looked for effective treatments to combat these problems and keep agricultural output at a steady level. Sulfur, copper, mercury, and arsenic were some of the first inorganic compounds used in pesticides. The synthetic pesticide industry's discovery of organic pest control chemicals in the second half of the 20th century was heralded as the "Green Revolution's" miracle cure to the world's expanding food demand and severe public health concerns like malaria and typhus. The pesticide industry has developed to the point that there are now more than 800 active chemicals accessible on the global market. There are 1,426 active pesticide compounds listed in the European Union's database, 476 of which have been authorized for use, 49 of which are pending approval, and 901 of which are either forbidden or have further constraints placed on their use. There was an increase in pesticide use across the board between the years 2000 and 2017; however, the rates of increase varied greatly amongst regions. Since 1990, the rate at which pesticide active ingredients are consumed per unit of agricultural land has increased by 75%, from 1.9 kg/ha to 3.3 kg/ha, as reported by the United Nations Environment Program (UNEP) in its report on Pesticides and Fertilizers. Increasing agricultural productivity is necessary to meet the rising demand for food in a world where both the population and the amount of arable land are on the rise. It has been suggested that more sustainable agricultural practices be employed due to worries about the potentially unsustainable use of pesticides and fertilizers to facilitate this intensification (Singh, Raj, & Yadav, 2022).

Soil can get contaminated with pesticides in several ways: by deposition from sprayed foliage, release from the treated surface of seeds, or application of granules or spray. Pipeline leaks, spills, damaged containers, trash dumps, and leaking underground storage tanks are all major causes of pesticide spread in agricultural soils. The packaging of pesticides, the failure to properly dispose of unused or expired pesticides, and the failure to thoroughly clean application equipment also contributes to contamination.

Acute and/or chronic effects on non-target organisms, including humans, have been documented for several pesticides with long half-lives. People started worrying about pesticide residues with the publication of the seminal book Silent Spring in the 1960s. Some of these chemicals can be degraded, leading to pesticide buildup in soils, groundwater seepage, and runoff transfer to surface water bodies. Pesticides can also be emitted into the air, carried by the wind, and deposited on land and water. How they behave in the soil and how they move to other mediums is determined by their chemical characteristics and half-lives, as well as by factors in the soil such as organic carbon content, texture, mineralogy, pH, and soil microorganisms (waterbodies, food, or organisms) (Singh, Raj, & Yadav, 2022).

Because of their ability to adapt to new environments, they pose a threat to human health and may easily enter the food chain.

Some pesticides, such as copper-based fungicides used in vineyards and orchards and pesticides and fungicides containing arsenic, copper, manganese, and zinc historically used to preserve citrus, apples, peaches, strawberries, and other fruit crops, can be sources of trace elements. Organotin chemicals, such as tributyltin oxide, were used in antifouling paints and were found in harbor sediments, whereas organo-mercury compounds, like *phenylmercury acetate*, were sprayed on seeds. Dredged sediments are often dumped in landfills or used as agricultural soil additions, however, these sites may or may not have leachate containment systems. A serious risk of soil contamination is posed by this. As a wood preservative, chromate copper arsenate was widely utilized. There are more than 1800 different active compounds listed in the Compendium of Pesticide Common Names, 219 of which include at least one trace element.

In response to growing concerns about the impact of an expanding number of potentially harmful chemical compounds on both human health and the environment, the international community ratified two chemical conventions around the turn of the century (including pesticides). In 2004, countries that had signed the Rotterdam Convention on the Use and Trade of Hazardous Chemicals took effect. The Stockholm Convention for the Control and Elimination of POPs was signed in 2001 and came into force in 2004. POPs (Persistent Organic Pollutants) persist for long periods and are dispersed widely due to natural processes including soil, water, and most importantly air. They endanger both people and animals by spreading through the food supply.

Figure 2.1. Many millions of hectares are sprayed from the air each year.

Source: Wang G, Lan Y, Yuan H, Qi H, Chen P, Ouyang F, Han Y. Comparison of Spray Deposition, Control Efficacy on Wheat Aphids and Working Efficiency in the Wheat Field of the Unmanned Aerial Vehicle with Boom Sprayer and Two Conventional Knapsack Sprayers. Applied Sciences. 2019; 9(2):218. https://doi.org/10.3390/app9020218The use of illegal or fake pesticides can have negative effects on ecosystems. Annually, between 5 and 15 percent of all pesticides sold worldwide are thought to be fake or otherwise unlawful. The use of illegal and counterfeit pesticides can pollute the soil and limit the viability of future harvests, on top of the possible harm to farmers' health, crop damage, and food safety (Ansari, A. A., Naeem, M., & Gill, S. S., 2018).

Pesticides account for a large proportion of the chemicals covered by both treaties; for example, 36 of the 53 compounds banned in Annex

III of the Rotterdam Convention, and 16 of the 26 substances that must be eradicated under Annex A of the Stockholm Convention are pesticides. Transfers of highly toxic pesticides from OECD countries to developing countries have declined marginally after the ratification of the Rotterdam and Stockholm Conventions (non-OECD). By the turn of 2000, however, many developing countries—particularly those in Africa, South America and the Caribbean, Central Asia, India, and Eastern Europe—held large stocks of antiquated pesticides. Because of the high cost of alternatives or the lack of availability, banning certain pesticides proved challenging. Unfortunately, pesticides are sometimes buried without proper containment or are kept in containers that have degraded through time and/or exposure to the elements. Most outmoded above-ground stocks have been collected and destroyed in an environmentally acceptable way thanks to the concerted efforts of the affected nations, the Global Environmental Fund (GEF), other donors, international organizations, NGOs, and the business sector. Pesticides that were buried have leaked and poisoned the ground, and the leaks have not been fixed (Singh, Raj, & Yadav, 2022).

Pesticide contamination and exposure take place in storage facilities, airport operations, points of use, and points of disposal for obsolete pesticides. However, a fresh stockpile of outmoded pesticides may be formed due to the ban on novel chemicals, inefficient pesticide procurement and stock management, and the emergence of counterfeit pesticides.

2.3 MINERAL FERTILIZERS

From soil solutions, plants receive nutrients and water through their root systems. These nutrients can be extracted from the soil by mineral dissolution, mineral desorption, soil organic matter breakdown, fungal hyphae, and microbial activity. Additionally, the soil may get nutrients from outside sources. Fertilizers of mineral, synthetic inorganic, and organic origin have made it possible to increase agricultural productivity across the world by supplying the necessary nutrients. Between 1960 and 2016, the worldwide population expanded by 145 percent, while the value of gross agricultural production increased by 250 percent. This enormous population growth would not have been possible without management techniques, such as nutrition provision. Between 1961 and 2002, phosphorus fertilization increased by 2.3 times, whereas nitrogen fertilization increased by 7.4 times. According to FAO research on worldwide trends in fertilizer use, fertilizer nutrient demand is expected to reach 199 million tons by the end of 2019, increasing at an annual rate of 1.6% (Ansari, Naeem, & Gill, 2018).

In some regions and nations, the excessive use of fertilizers has caused severe environmental issues, such as nitrogen saturation in soils, fertilizer loss by leaching into groundwater, discharge into surface water, and eutrophication of freshwater rivers, lakes, and oceans. There are several reasons for this excessive use, including:

- Low cost in comparison to what is seen to be a low danger of misuse in comparison to diminished yields;
- Lack of product knowledge and teaching.
- Lack of current and sufficient fertilizer recommendations for a variety of crops, especially tree crops.

To fulfill the expanding global food demand, address the limited availability of critical nutrients such as phosphorus, and alleviate environmental problems caused by excessive fertilizer use, it is necessary to boost fertilizer usage efficiency (e.g., eutrophication). FAO developed a Code of Conduct for Sustainable Usage and Management of Fertilizers in response to the need for more sustainable fertilizer usage. The code provides a mechanism for implementing the Voluntary Guidelines for Sustainable Soil Management concerning nutrient imbalances and soil contamination. The code endorses techniques that increase soil health, such as nutrient recycling and agronomic and land management. As part of its preparations for the common agricultural policy for 2021–2027, the European Union is creating a new tool to assist farmers in managing the usage of nutrients on their farms. The Farm Sustainability Tool for Nutrients seeks to accelerate agricultural automation while enhancing the sustainability of fertilizer usage across all European Union farms (Ansari, Naeem, & Gill, 2018).

Nitrogen is lost from the soil owing to volatilization, denitrification, leaching to groundwater, surface runoff, and erosion when nitrogen fertilizer is overused. Depending on the kind of nitrogen and soil conditions, microbial activity converts nitrogen into nitrous oxide (N2O), one of the greenhouse gases that contribute to global warming and the stratospheric ozone hole. Different nitrogen fertilizers contribute to soil acidity, acid rain, and eutrophication through a range of soil and environmental transformation mechanisms. Nitrogen-containing organisms interacting in the atmosphere can generate aerosol particles that are hazardous to both the environment and human health. According to recent studies, soil acidification is an environmental condition that can reduce the number of plant nutrients available in the soil for plant growth, cause direct toxic effects from soil elements such as aluminum, increase the risk of soil-borne plant diseases, and diminish a plant's ability to defend itself against pathogen attack.

Phosphorus in the soil is depleted by crop uptake and erosion, while it is slowly replenished by rock weathering and air deposition. As a result, the use of both inorganic and organic phosphorus is currently widespread in crop production. In modern agriculture, phosphate generated from phosphate rock (PR) is mostly derived from sedimentary ores (75–80%) and is therefore a nonrenewable resource. Phosphate rock contains trace elements and radionuclides in varying concentrations, particularly in sedimentary ores. Trace elements and radionuclides are mobilized during the phosphate mining process, and the resulting dust is a source of fluoride and radon emissions. When phosphate fertilizer is applied, hazardous compounds such as arsenic, cadmium, chromium, lead, mercury, fluorine, and radionuclides such as uranium, radium, and thorium are added to the soil, where they can accumulate in plants and enter the food chain. Multiple applications of fertilizer might result in a considerable buildup of potentially hazardous substances in the soil. In soils with a lengthy history of phosphorus fertilizer application, higher levels of copper, zinc, and cadmium were detected (Ansari, Naeem, & Gill, 2018).

Figure 2.2. Fertilizer application on plants.

Source: Xiao Q, Du R, Yang L, Han X, Zhao S, Zhang G, Fu W, Wang G, Lan Y. Comparison of Droplet Deposition Control Efficacy on Phytophthora cap-

sica and Aphids in the Processing Pepper Field of the Unmanned Aerial Vehicle and Knapsack Sprayer. *Agronomy. 2020; 10(2):215. https://doi.org/10.3390/agronomy10020215 The world's known phosphate rock reserves are dwindling, and a 0.7–1.3% yearly rise in consumption would likely lead phosphorus demand to exceed supply by 2035. The International Fertilizer Development Center's research of global phosphate deposits and resources, as well as the possibility for more efficient extraction and processing technologies, revealed that reserves can survive for a significantly longer period.*

The presence of cadmium in phosphate rock derived from 35 sedimentary sources in 20 nations. Globally, the average cadmium level of sedimentary phosphate rock formations is approximately 69 times greater than that of non-phosphate-containing rocks. Cd levels may vary from less than 1 mg/kg to 150 mg/kg between countries and deposits within the same country. The largest quantities have been discovered in Tobene, Senegal (60 - 115 mg/kg) and Idaho, United States (40 - 150 mg/kg) phosphate rock. Approximately 85 percent of the world's fertilizer supply is derived from sedimentary deposits. The average cadmium concentration in igneous phosphate rock is about 2 mg/kg. Cadmium is transferred from the manufacturing process to the final product. In contrast, scientific risk assessments indicate that phosphorus fertilizer containing cadmium is safe and poses no hazard to human health (Menjivar et al, 2018).

If an excessive amount of phosphorus fertilizers is utilized, croplands may lose phosphorus through runoff or erosion. The eutrophication of nearby water bodies is caused by excessive phosphorus loss from agricultural fields as a result of erosion or runoff. Phosphorus losses in eroding agricultural soils are especially difficult to assess because erosion rates vary greatly even within a single area. It is largely attributable to the fact that few nations conduct extensive and frequent soil erosion studies.

As a byproduct of the production of phosphate fertilizers using wet acid extraction, phosphor-gypsum is produced. Annual production of 100-280 Mt of phosphor-gypsum is expected. Phosphor-gypsum can be enriched in radon, uranium, and thorium, depending on the phosphate rock source. It also includes trace elements such as barium, cadmium, copper, nickel, strontium, and zinc in varying concentrations. Because of this, the United States has categorized waste from the manufacturing of phosphate fertilizer as naturally occurring radioactive material that has been technologically improved (TENORM). Under ambient environmental circumstances, leachates from these stockpiles demonstrated enrichment in trace elements connected to the organic component of the phosphor-gypsum waste and mobilization of U from the crystalline structure in the original phosphate rock to the bioavailable fraction in the trash. Typically, phosphorus fertilizer manufacturing wastes end up in open heaps.

In addition, phosphor-gypsum, a byproduct of the production of high-analysis phosphate fertilizers using phosphoric acid, has been utilized as a soil amendment in several nations for a long time to improve the soil's characteristics and crop yields. Phosphor-gypsum has tremendous potential for liming and trace element immobilization because of its high calcium concentration. However, there is evidence of radionuclide and trace element accumulation in soils treated with phosphor-gypsum, pyrites, and other amendments, depending on their source. Monitoring the amount of these pollutants in soils and their entrance into the food chain is vital, since supporting such recycling might enhance phosphor-gypsum use (Menjivar et al, 2018).

2.4 ORGANIC FERTILIZERS

In agriculture, animal dung, compost, septic sludge, sewage sludge, municipal biosolids, and food industry wastes are common organic fertilizers. The use of organic fertilizer improves soil health by delaying the nutrient release and enhancing organic carbon content. Due to the mineralization of organic nitrogen, higher nitrate concentrations, the presence of trace elements, perfluorinated alkylated substances (PFASs), brominated flame retardants, and other hazardous compounds, it may become a cause of soil contamination.

The usage of organic fertilizers has increased due to the growing interest in organic agricultural techniques and the demand for organic food. Recycling organic waste in agriculture is considered an ecologically responsible and sustainable disposal strategy.

Manure and animal wastes

Depending on the quantity of total solids present and how the manure is collected, stored, transported, and applied to the fields, animal manure management methods can be characterized as solid, slurry, lagoon systems, or a combination of these. Through anaerobic digestion or composting, animal waste may be pre-treated. Through anaerobic digestion, methane is created, which may be utilized to generate energy or heat fields.

In addition to essential minerals such as copper and zinc, feeds used in intensive livestock production may also include chromium, lead, and other potentially hazardous metals. Adding necessary trace elements to animal feed is a common method of increasing output. Researchers discovered an increase in the amounts of trace elements in manures between 1990 and 2010. Copper and zinc consumption in animal feed increases copper and zinc concentrations in manure, hence harming agricultural soils. Various trace elements are utilized in animal production to meet certain dietary requirements. Copper sulfate is used as a growth stimulant in pig and poultry systems and as a footbath in dairy cattle to cure lameness. As a therapy for scour, zinc oxide is given to pig food.

The prevalence of antimicrobials and antimicrobial-resistant organisms in soils that have been treated with organic fertilizer of human or animal origin is on the rise. Antibiotics are commonly supplied prophylactically to animals, not just to prevent disease but also to enhance growth. 75 to 90 percent of prescribed antimicrobials are not fully metabolized in the systems of humans and animals, are excreted, and accumulate in animal manure, sanitary wastewater, wastewater sludge, and biosolids, according to reports.

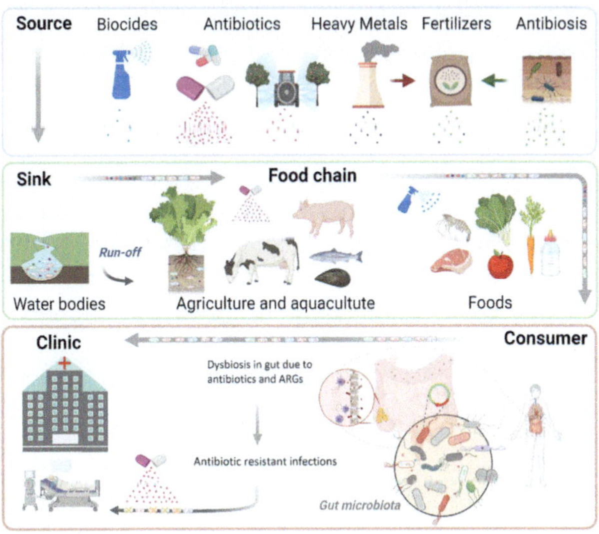

Figure 2.3. Antimicrobials in soil.

Source: Skandalis N, Maeusli M, Papafotis D, Miller S, Lee B, Theologidis I, Luna B. Environmental Spread of Antibiotic Resistance. Antibiotics. 2021; 10(6):640. https://doi.org/10.3390/antibiotics10060640 Low antimicrobial *concentrations in soil can result in the spread of antimicrobial resistance due to modifications in the bacterial genome and the transmission of antibiotic resistance genes (ARGs) and associated mobile genetic elements (MGEs) within and across microbial populations. Consequently, the overall sensitivity of microorganisms to antimicrobials changes. Antibiotic resistance is becoming recognized as a significant threat to human health (*Đolić *et al, 2022).*

Sewage sludge and biosolids

Due to the diversity in wastewater quality and treatment methods, several semi-solid and solid wastes that can be utilized as organic fertilizers are generated. These byproducts include septic sludge, which is untreated septic tank waste, and wastewater sludge which has undergone primary and secondary treatment, such as biological digestion and settling. Significant quantities of biosolids are produced and utilized to enhance agricultural land (for example, in Japan, China, and the United States of America, respectively, 70, 30, and 6 million tons). Even though biosolids have been examined and determined to be suitable for this purpose, the safety of utilizing sewage sludge as fertilizer is a concern. Wastewater comprises two distinct categories of contaminants: hydrophilic pollutants that stay in the treated aqueous fraction and lipophilic pollutants that interact with the solid fraction.

Due to their high concentration of organic matter and biogenic components, biosolids may include substantial amounts of plant growth nutrients such as nitrogen and phosphorus. As a byproduct of wastewater treatment facilities, the use of biosolids as fertilizer in agricultural regions is an excellent approach to improve the organic matter and nutrient content of the soil while minimizing waste. Biosolids are a substantial source of trace elements when employed as a soil amendment; however, they are also contaminated with lipophilic trace elements. "Biosolids" refers to both micro- and nano-scale polymers. Trace elements and POPs, such as *perfluorinated alkyl substances (PFASs)*, such as *perfluorooctane sulfonate, polychlorinated biphenyls (PCBs), chlorinated paraffin (CPs), and polychlorinated dibenzo-p-dioxins and dibenzofurans (PCDD/F),* may be present in wastewater sludge and biosolids (PFOS). However, whereas PCDD/F and PCB levels in compost and sewage sludge have decreased over the past few decades, levels of other persistent organic pollutants such as PFOS/PFOA and CPs have increased. Moisturizers, lipstick, shampoos, hair colors, and deodorants are some examples of pharmaceuticals and personal care items that enter the sewage system after usage. Analgesics, antibiotics, antiepileptics, -blockers, blood-lipid regulators, hormones, antiretrovirals, and steroids were identified in wastewater and sludge. Several of these drugs are also known to impact the body's hormones.

Compost

Food and green waste from families, towns, and enterprises may be fermented and composted into organic fertilizers. Recent investigations have discovered that this form of fertilizer contains microplastic particles that are difficult to remove. Compost created from municipal solid waste has been identified as a source of trace elements when used as an organic fertilizer on agricultural soils.

Using just manures and other organic fertilizers with their NPK ratios might result in an excess of one of the components. Excessive phosphorus application can result in leaching and ultimately eutrophication of freshwater bodies of water. Excessive nitrogen application can result in nitrogen leaching, ammonia volatilization, nitrous oxide emissions, and eventually eutrophication of marine environments. It is essential to utilize the correct NPK ratio when applying organic or synthetic fertilizer to avoid harmful impacts. In China, nitrate exceeded the WHO drinking water limit of 11.3 mg/L N in 55% of 394 groundwater samples.

2.5 WASTEWATER FOR IRRIGATION

There are various types and levels of pollutants in wastewater, depending on the source of the effluents and the treatment procedures employed. In developing nations, wastewater is frequently utilized for irrigation, especially in or near cities where it is abundant. This is especially prevalent in or near cities where it is abundant. According to the most recent statistics, approximately 0.31 109 m3/year of treated municipal wastewater is utilized for irrigation globally. The utilization of this affordable irrigation supply aids urban and peri-urban agriculture, while the organic contaminants reduce the need for chemical fertilizer. On the other hand, untreated wastewater irrigation can lead to the buildup of trace elements, and organic pollutants such as POPs and PFAS, and be a source of harmful microorganisms, pharmaceuticals, and personal care items.

Few nations have reported utilizing untreated wastewater for direct irrigation; the volume varies widely, reaching 4 million cubic meters per year in Mexico. Due to the possibility of pathogen infection and trace element absorption and buildup in edible crop components, consuming food that has been irrigated with untreated wastewater may constitute a significant health risk. Moreover, untreated sewage includes microplastics. Untreated wastewater irrigation can also have a detrimental impact on the soil's physical properties, such as the development of water repellency (Chauhan A, 2015).

Secondary and tertiary treated wastewater irrigation is less prevalent in more affluent nations, although it is highly common in water-scarce regions such as China, Mexico, the Middle East, and North Africa. In Israel, almost fifty percent of agricultural productivity is derived by irrigation utilizing treated wastewater. Due to the common presence of pollutants in residual biosolids, treated wastewater contains low concentrations of contaminants and microplastics.

Before the Stockholm Convention identified *perfluoroalkyl substances as POPs, perfluoroalkyl* (PFAS) compounds were often used in firefighting foams. Their poisoning of irrigation and groundwater has become a significant problem.

2.6 AGRICULTURAL PLASTIC WASTE

The usage of plastic in agriculture has expanded in recent decades. In 2019, 56% of the 708 000 tons of non-packaging plastic used in agriculture in the European Union were used to produce silage and twine for baling cattle feed. The remaining 44% was allocated to greenhouse films, mulching films, micro tunnels, drippers, irrigation pipes, and crop protection fleeces and nets. Plastic is utilized in the production of seed and fertilizer bags, pesticide bottles, seedling trays, and containers. Many of these items are intended for single-user use. In the absence of effective procedures for their recovery from fields, collection, and ecologically responsible recycling or disposal after use, agricultural plastics are a major cause of soil contamination and environmental pollution. The government and scientific community just recently became aware of plastic contamination in agricultural regions.

Mulching films

Mulching films have the largest potential for soil contamination of any agricultural plastic product due to their proximity to the soil and the possibility of soil absorption during decomposition or post-harvest treatment. Plastic mulch is applied to boost crop yield by protecting seedlings, reducing evaporation, and reducing the impact of weeds and insects. It is commonly utilized with the plastic drip irrigation tubing below. On crops that rotate, a mulching film is typically used for only one planting season. Between 1991 and 2011, China's use of plastic film mulching climbed by a factor of four. China is the world's largest consumer of plastic film mulching. In 2011, China was said to have covered 20 million hectares of land with plastic film, whereas, in Europe, 427 000 hectares of agricultural land were covered with plastic mulch film. It is projected that the area covered by mulching has expanded, despite the absence of recent data due to prior growth trends.

1978's introduction of standard polyethylene plastic mulch films is the primary source of plastic in agricultural soil. It is desirable because of its adaptability and durability, but recycling and field recovery are challenging and costly operations. Polyethylene has been the most frequently used polymer for decades, but new biodegradable plastic films are being developed to replace it to decrease post-harvest cleaning expenditures. Regarding their total biodegradation and the influence of byproducts and additives (such as plasticizers) on soil organisms and structure throughout the biodegradation process, there are still many unknowns. Instead of entirely dissolving, plastics are fragmented into ever-smaller particles (micro- and nano-plastics) that can enter the food chain. Because animals that consume soil particles might absorb them and transport them to the food chain, the presence of micro- and nano-plastics in soil poses significant problems. These obstacles might make selecting these options difficult. Oxy-degradable plastics and biodegradable bioplastics are the two basic types. Oxy-degradable polymers are standard synthetic plastics such as polyethylene that have additives that promote film oxidation and fragmentation. Concerned that these films contribute to micro-plastic contamination, several European countries have banned their usage. After harvest, biodegradable bioplastic mulching films are thought to decompose in the soil and may be safely tilled in (Andrews, Karlen, & Cambardella, 2004).

The widespread use of traditional plastic mulching, particularly in China, has resulted in enormous accumulations of plastic garbage in the soil and a decline in agricultural yields. Micro-plastics are fragments smaller than 5 mm, and if they continue to degrade, they will likely transform into nano-plastics, which are more difficult to detect. Plastic mulch pollution has a detrimental influence on plant growth by affecting soil structure, seed germination, microbial activity and community, and root development. It has the potential to enhance greenhouse gas emissions. Another issue is that the remaining plastic mulch includes chemicals, some of which are carcinogenic and hormone disruptive. They are the primary source of phthalates in soils, but they are also capable of absorbing other very hazardous chemical compounds, which may be released when they are changed or come into contact with live organisms. Despite the need for more research to fully comprehend the issue, their release into the soil poses a significant threat to both the ecosystem and human health.

Greenhouse film and protective netting

All continents rely heavily on greenhouses to provide food throughout the year. Europe has around 175,000 hectares of protected horticulture zones in 2015, whereas China had almost 3,300,000 hectares of greenhouse-covered land. In industrialized nations, glass greenhouses are replacing plastic greenhouses, while plastic is still utilized for some greenhouse applications and has a low recycling rate, resulting in plastic waste. In underdeveloped nations, plastic films continue to be the most common inexpensive solution. Although protective nets and greenhouse films are designed to last several harvests, weathering and photo-degradation ultimately necessitate replacement, with the old materials recycled or disposed of appropriately.

Other agricultural plastic wastes

As was previously indicated, a lot of additional agricultural plastic goods wind up in the garbage on farms and, if not collected and disposed of correctly, may cause soil contamination. To prevent their incorporation into the soil during post-harvest management, non-biodegradable mulching films, irrigation tubing, micro tunnel films, and protective fleeces must be removed from the fields with care and precision. After planting, polystyrene seedling trays must be gathered and removed from the fields. Gather and store used silage and bale films, netting, and twine. Due to pesticide residue, empty pesticide containers are extremely harmful. Pesticides for small-scale farmers are often offered in single-dose sachets in less developed nations. A particular concern is a collection and proper disposal of contaminated sachets discovered lying in fields. Technical recommendations for the ethical handling of empty pesticide containers are added to the International Code of Conduct on Pesticide Management. Empty seed packets should be handled with care since they may contain pesticides.

Multiple nations demand that producers and their extended supply chains collect plastic waste from farms and dispose of or recycle it in an ecologically appropriate manner. Many of these organizations began by giving free help for empty pesticide containers, but have now expanded to handle a larger variety of agricultural plastic garbage. For instance, such schemes exist in France and Brazil for PEVs. In the absence of such collecting systems, there is a risk of soil deterioration if plastic trash is unlawfully disposed of on a farm. Under its EPI-Agri program, the European Commission has formed a focus group to investigate the decrease in the environmental impact of agricultural plastics. The Stockholm Convention regulates POPs, which include polychlorinated dibenzo-p-dioxins and polychlorinated dibenzofurans.

2.7 RURAL COMMUNITY WASTE

Similar forms of waste are generated in rural regions, but in lesser amounts, making it more challenging and costly to collect, recycle, or dispose of in an ecologically responsible manner. If waste is not collected for proper environmental management, it is likely to be thrown at an uncontrolled open dump or carelessly discarded in the countryside. Garbage may be burned outside to minimize volume, and uncontrolled burning emits toxic pollutants, such as partially burned particles containing contaminants such as PCDDs and PCDFs, which can have a short- and long-term impact on soils.

2.8 URBAN AND PERI-URBAN AGRICULTURE

Because of the rise in urban population and food demand, urban and peri-urban agriculture is expanding quickly. FAO estimates that around 800 million people worldwide cultivate food in urban gardens, which has a significant positive influence on household incomes, food security, and nutrition, as well as providing urban green space and climate resilience. However, if the soil is contaminated, the risk to human health and the possible transmission of toxins from urban soils to crops require additional consideration.

Urban horticultural gardens and parks may include trace elements and persistent organic pollutants (organochlorine pesticides [OCPs], PCBs, PCDD/Fs, PBDEs, and PAHs). Due to the bioaccumulation of these pollutants in poultry and livestock as well as repeated contact with contaminated soils during recreational activities — which is especially hazardous for young children — they may pose a threat to human health. Due to high trace element concentrations in lettuce and potatoes cultivated in household gardens in southern Sweden, a historically extensively polluted region by the glass production sector, daily cadmium intakes surpassed the WHO's maximum acceptable intake levels. Similar outcomes were observed with cadmium and lead accumulation in lettuce grown in French household gardens, but with the less human application. The concentration of trace elements in the soil cannot predict plant absorption and the possible risk to human health due to the involvement of other variables, including soil parameters, plant traits, and plant metabolism. However, the amount and proportion of bio-available trace elements in soils must be considered when evaluating hazards. Additionally, it is prudent to monitor persons who consume products from polluted urban gardens due to the possibility of long-term cumulative effects from food consumption and direct contact with contaminated soil particles.

Local retailers provide synthetic insecticides and fertilizers for residential and small-scale applications. National chemical regulation and enforcement, as well as the incidence of illicit and counterfeit goods, will have an impact on the variety and quality of market-available items. The high quantities of lead, chromium, arsenic, and cadmium in the soil of both public and private gardens in New York City pose a major risk to human health. Commonly, agrochemicals are over applied due to insufficient knowledge of the optimal rates and application methodologies, resulting in soil buildup. Urban gardening plots share the same environmental issues as rural regions, such as the excessive use of pesticides, irrigation with impure water or recycled water, and the usage of contaminated home compost.

2.9 ACCIDENTAL LEAKAGES

Accidents and unintentional leaks are other big contributors to urban soil contamination. Landfills, sewers, septic tanks, fuel tanks, and pipes all contain high quantities of a wide variety of pollutants. PCBs may be released from power transmission equipment. All of them are susceptible to unintentional or incidental losses as a result of natural catastrophes such as hurricanes, earthquakes, or fires, as well as inadequate maintenance leading to degradation and eventual rupture and leaking, and as a result of exterior damage during excavation.

Although they can also be caused by combustion emissions, there is a strong association between oil spills and the occurrence of PAHs in urban soils. During the analysis of PAHs in Shanghai urban soil, six sources of the total soil PAH load were identified: petrogenic sources (6%), coal combustion (21%), biomass burning (13%), creosotes (16%), coke manufacturing (22%), and vehicle emissions (21 percent). The presence of

PAHs adjacent to research locations in Antarctica that would otherwise be essentially uncontaminated demonstrates the connection between oil spills and the occurrence of PAHs in the environment. Due to their frequency and potential consequences on human health and the environment, preventing oil spills should be a key priority on a global scale, as should enhancing regional collaboration. The quantity of oil in soil has a substantial effect on plant development and metabolism. While the germination rates of two acacia species were significantly reduced, the gas exchange and photosynthetic rates of legumes growing in oil-polluted soil declined. Changes in soil microbial communities have been reported when oil-degrading bacteria outnumber native species in soils contaminated with oil.

Leakage from underground gasoline tanks at gas stations for automobiles is also prevalent, with the contaminated plume moving through the soil and groundwater. In 1984, the U.S. government initiated a program to repair these leaky subterranean storage tanks. By 2010, 80 percent of the 588 000 underground storage tank leaks had been repaired (Varjani, & Upasani, 2019).

Sewer lines and septic tanks connected to home and hospital wastewater discharges exhibit significant quantities of PPCPs, fecal bacteria and other pathogens, antimicrobial-resistant bacteria, micro-, and nano plastics, and poly- and perfluoroalkyl compounds, in addition to POPs and trace elements. Diverse toxins will be discharged into the earth from a sewer system break, with the possibility for them to reach groundwater. New pollutants, such as PPCPs, are now being studied for their potential effects on human health and the environment. In addition, research on pollutants in wastewater has paid little attention to how contaminants are kept in the soil matrix; instead, they have mostly concentrated on identifying contaminants in ground and surface water bodies.

2.10 ELECTRONIC WASTE

The category of electronic garbage, or e-waste, includes electronic devices such as computers, mobile phones, and home appliances. As their rate of growth quickens and they become more inexpensive, the operational life of high-tech items decreases as they are replaced more frequently. As a result, e-waste output has surged dramatically, reaching 44.7 million metric tons in 2016, and is anticipated to climb by about 14% by 2021 (Siddiqa, A., & Faisal, M, 2020). E-waste is a rich and diverse source of materials, including rich and rare minerals such as gold, copper, neodymium, and indium, trace elements such as cadmium, chromium, lead, and mercury, and persistent and emerging pollutants such as POPs, PCBs, plastics, perfluorooctanoic acids (PFOAs), and flame retardants.

Numerous high-income nations have enacted legislation to boost the rate of e-waste collection and ecologically friendly handling. Typically, these rules are founded on extended producer responsibility. For economic reasons, however, a substantial amount of e-waste was transferred to underdeveloped nations. It is estimated that only 20% of worldwide e-waste is effectively collected and processed, with the remaining 80% ending up in urban dumps or landfills, primarily in China, India, Mexico, Brazil, Eastern Europe, Central Asia, and West Africa. Receiving nations generally have less stringent e-waste restrictions, and rubbish is regularly stored in uncontrolled squatter sites. A huge number of urban poor people who make a living by deconstructing obsolete equipment for reuse or recycling it to recover rare minerals and materials are exposed to hazardous pollutants found in garbage and dirt, which disproportionately affects women and children. Even though data on e-waste recycling exclude informal recycling and reassembly, it is a significant activity in receiving countries and one of the primary sources of soil pollutants connected with waste streams in these nations. Since the Basel Convention's Ban Amendment, which prevents industrialized nations from exporting hazardous waste to developing countries, entered into effect on December 5, 2019, significant progress has been achieved in decreasing pollution in recipient countries.

In China, the prevalence of pollutants in soils near uncontrolled e-waste recycling and processing enterprises has been the subject of several investigations. Before the implementation of the ban, China was the leading importer of e-waste. It is projected that e-waste recycling facilities in China and India would become the primary source of highly contaminated regions as a result of increasing home use. Extremely high amounts of brominated flame retardants, PAHs, PCBs, and trace elements were detected in the soil at an e-waste open-burning site in southern China. Except for volatile chemicals, the majority of concentrations were within 500 meters of the e-waste open-burning site (PAHs) (Siddiqa, A., & Faisal, M, 2020).

Agrochemicals

In recent decades, crop protection has expanded significantly. The majority of industrialized nations regulate and monitor the production of agrochemicals to minimize their detrimental consequences. Manufacturing generates significant amounts of waste and contaminants, which must be managed utilizing the most recent technology. In countries with fewer regulations, these wastes and toxins still represent a threat to the soil's health.

The pesticide manufacturing business has abandoned hazardous waste generated as a byproduct of the production process. Extremely hazardous organic pollutants such as 1, 2, 3, 4, 5, and 6-hexachlorocyclohexane (HCH), DDT, and chlordane have been connected to the production of organochlorine pesticides. Several of these pesticides have been prohibited because of their environmental persistence and propensity to cause cancer and/or neurotoxicity. However, soils near production and application areas continue to contain residues. One of these legacy difficulties is the production of the organochlorine insecticide Lindane, the gamma isomer of *hexachlorocyclohexane*, which was previously one of the most frequently used pesticides in the world. Approximately 6–10 tons of byproducts, primarily alpha and beta HCH, are created for every tone of active Lindane manufactured. These byproducts, which are persistent and bio-accumulative and have the same toxicity and environmental impacts as Lindane, were added to the Stockholm Convention as POPs in 2017 due to their POP status. Only pharmaceutical usage against head lice and scabies is permitted; the manufacture of Lindane has been banned. Significant environmental problems were produced by the buildup of leftovers of Lindane production in open stockpiles and landfills, some of which lacked adequate containment mechanisms (Dantu et al, (2020).

Chapter

3

IMPACTS OF AGRICULTURAL CONTAMINANTS ON BOTH PEOPLE AND ANIMALS

Contents

3.1 Introduction..44

3.2 Farmers and their Contribution to the Effects of Pesticides on Human Health..45

3.3 Pesticides Associated with Health Risk Factors..46

3.4 Biomagnification and Bioaccumulation of Contaminants in Aquatic Animals..54

3.1 INTRODUCTION

As the world's population grew, industrial farming appeared to be the obvious solution. Scientists promoted high-yielding cereal hybrids, chemical pesticides, and synthetic fertilizers to eradicate hunger, feed high-growing populations, and revive flagging economies. Between the early 1960s and 2015, agricultural output more than tripled in intensity, producing an abundance of cheap food and averting a global food shortage.

However, things did not go as smoothly as planned. Decades of industrial farming have had disastrous consequences for the natural world, and the prospects for long-term food production are bleak at best.

Most developed and developing countries now blame pollution from farms and ranches rather than cities and factories for degrading their coastal and inland water supplies. According to the World Water Assessment Program (2013), agricultural nitrate is the most common chemical contaminant in groundwater aquifers worldwide. According to WWAP (2015), agricultural pollution significantly impacts 38% of EU water bodies. Agriculture is the second largest contributor to wetland pollution in the United States and the third largest to lake pollution. The agricultural sector is largely responsible for nitrogen pollution in China's groundwater and much of the country's surface water (FAO, 2013). In low- and middle-income countries, untreated municipal and industrial wastewater is a major issue.

On the other hand, sediment runoff and groundwater salinization exacerbate an already serious problem: agricultural pollution. Cropping, livestock, and aquaculture systems have all expanded and intensified to meet rising food demand caused by population growth and dietary shifts. According to FAO reports from 2014 and 2016, the total number of livestock has tripled in recent decades, and the amount of irrigated land has more than doubled. Since the 1980s, there has been a more than 20-fold increase in aquaculture, particularly in inland and Asia.

Crop production increases worldwide can be attributed largely to the widespread use of inputs such as pesticides and chemical fertilizers. This pattern has been accelerated by the increase in arable land, with irrigation increasing output and rural incomes while dispersing agricultural pollution into waterways.

3.2 FARMERS AND THEIR CONTRIBUTION TO THE EFFECTS OF PESTICIDES ON HUMAN HEALTH

Farmers cannot grow healthy, contaminant-free food without the use of pesticides. Without them, many of their food crops would be destroyed, resulting in waste and higher grocery and restaurant prices.

Pesticides and synthetic and modern antimicrobials are far superior to their historical counterparts. Unfortunately, pesticides are frequently misused, posing serious health risks and even death. Concerns about agricultural pesticides' toxicological risks to humans and the environment are growing, particularly in developing countries (Menjivar et al, J, 2018).

The following are some of the factors that contribute to an increase in the use of poisonous pesticides that harm human health:

- In most of the country's agricultural regions, pesticide misuse is exacerbated by illiteracy. Farmers who are unfamiliar with agriculture frequently seek advice or inspiration from their neighbors to begin growing crops. Furthermore, because illiterates can work for less money, some farmers may choose to hire them. Pesticide labeling, for example, maybe too complicated for farmers to understand.
- Unsafe and unrecommended practices: When it comes to pesticides, many farmers disregard the instructions. For example, some farmers believe applying pesticides at higher rates than recommended will increase their efficacy. Some farmers mix pesticides with their bare hands, while others reuse empty containers around the house.
- Most farmers do not have secure storage areas for pesticides on their properties. Pesticides should be spread around the farm rather than kept in secure containers.
- Fourth, failing to read the directions. Most farmers fail to read the pesticide label instructions, including when and how to harvest.
- Inadequate pesticide training and storage Pesticide education and training for farmers are generally lacking, and this causes them to behave in inappropriate ways that they justify to themselves.
- Personal protective equipment (PPE) such as aprons, gloves, gumboots, and masks are rarely used by farmers. When it comes to pesticides, they rarely follow proper safety procedures.
- Even though product sales are their primary focus, agrochemical

companies play an important role in disseminating useful agricultural information. However, most of them do not, and among those who do, representatives tasked with assisting farmers are frequently unwilling to do so.

- When pesticides are used incorrectly, they pose serious health risks. Pesticide exposure can cause various symptoms, including headaches, excessive salivation, nausea, diarrhea, respiratory depression, seizures, and loss of consciousness. Proper pesticide safety procedures must be adopted and implemented to avoid unintended consequences. Simply put, the desire to increase agricultural output must coexist with the desire to protect people, food, and the environment.

How can this pollution be prevented?

- To improve farmers' knowledge and skills, the government should sponsor pesticide safety education (both formal and informal) and training programs.
- Before using pesticides, farmers, especially those with higher education, should read and completely comprehend the pesticide application instructions.
- If a farmer employs someone to tend to their crops, the hired help must have the farmer present when any pesticides are applied (if they must employ illiterate persons). They also seek to teach students about effective pesticide management procedures, focusing on pesticide storage and post-harvest monitoring.
- Agrochemical businesses can assist farmers in various ways, including education, exposure to cutting-edge technology, and encouragement to reject unsuccessful approaches in favor of more successful ones.

3.3 PESTICIDES ASSOCIATED WITH HEALTH RISK FACTORS

Pesticides are compounds or mixtures of substances used primarily in agriculture or public health programs to protect plants from pests, weeds, or diseases or humans from vector-borne diseases such as malaria, dengue fever, and schistosomiasis. This category includes chemicals that kill insects, fungi, herbs, rodents, and even plants. These pesticides can also maintain and develop non-agricultural areas such as public urban green spaces and sports fields. Furthermore, these chemicals eliminate or prevent pests in places like boat bottoms, construction materials, and pet shampoos.

Because of the risks they pose to human health and the environment, many pesticides are no longer used. Pesticides can be applied topically, inhaled, or consumed. Numerous factors influence potential health effects, including the type of pesticide, the amount exposed, the route of exposure, and the individual's current health status. Pesticides, whether human or animal, can be metabolized, excreted, stored, or bioaccumulated in adipose tissue. Pesticides have been linked to various health issues, including skin, digestive tract, nervous system, lungs, reproductive system, and endocrine system problems. Excessive pesticide exposure, whether intentional or unintentional, is hazardous and can be fatal (Menjivar et al, J, 2018).

Pesticide residues have been discovered in many foods and drinks that people consume daily, including pre-made meals, water, wine, fruit juices, snacks, and even animal feeds. It's also important to understand that even thorough washing and peeling will not remove all residues. The majority of concentrations are well below the legal limits. However, these "safe limits" may be too low, especially when multiple chemicals are present simultaneously and can have additive or synergistic effects, both of which are common in the real world. The discovery of pesticide residues in human breast milk samples raises concerns about prenatal exposure and child health.

The following are some of the common classes of pesticides causing health risk factors

Organochlorine Pesticides

Dichlorodiphenyltrichloroethane, or DDT, is the most well-known organochlorine insecticide. Its indiscriminate use caused a slew of issues for public and environmental health. Organochlorines used in insecticides include methoxychlor, dieldrin, endosulfan, heptachlor, and dicofol.

Some countries continue to use or plan to use DDT to protect their citizens' health. Like many other chemicals, DDT must be dissolved in various solvents before use. A molecule is considered present in all living things on Earth. DDT and its metabolite, p,p-dichlorodiphenyldichloroethylene (DDE), have also been shown to harm the endocrine system and cause cancer. There is evidence that DDT and DDE exposure during pregnancy can harm a child's brain development.

Organochlorine insecticides have been linked to various negative health effects, including hematological and hepatic changes, effects on embryonic development, changes in lipid metabolism, and endocrine abnormalities. Although their carcinogenicity is debatable, concerns about their potential impact on human health should not be dismissed.

Case study on DDT/DDE

Millions of infections and deaths caused by diseases transmitted by insects could have been avoided in the past if DDT had been utilized. Recent study show DDT and DDE are ingested by humans; indoor residual spraying can result in considerable exposure, and DDT may constitute a risk to human populations. However, very few studies have explored the effects of DDT/DDE exposure at levels seen in people exposed to indoor residual spraying. Studies have rarely assessed DDE and DDT body burdens. Furthermore, there is limited evidence on DDT/DDE exposure when repercussions such as cancer are most likely.

International policymakers should take into account the following issues, given the continuously expanding understanding of the impacts of DDT usage on humans:

- People in the United States who live near DDT production facilities or work in occupational settings have been exposed to DDT. If biomonitoring reveals chronically elevated levels of DDT or DDE, state and federal agencies should monitor residents and conduct a health effect study.
- Few studies on the health effects of residual DDT spraying have been conducted in locations where DDT is sprayed inside. These populations may differ from previously studied in ways that modify susceptibility and are likely to have received far higher DDT doses. To determine the exposure and health risks associated with DDT used for indoor residual spraying in the relevant communities.
- People with weakened immune systems, pregnant women, and children may be especially prone to the adverse effects of DDT. HIV/AIDS infection rates are high in many DDT-treated areas where malaria is prevalent.
- Breastfeeding is the most significant source of nourishment for newborns for up to one year, according to the American Academy of Pediatrics (2005). Some African countries allow women to breastfeed for up to two years. Due to the lipophilic nature of the chemicals, breast milk is a substantial source of DDT/DDE

exposure for neonates. If breastfeeding is avoided due to high DDT exposure, significant public health consequences may result.

- DDT may be a beneficial temporary malaria control strategy, but human exposure to this chemical should be limited. DDT exposure could be reduced by rigorously adhering to the criteria for indoor residual spraying, informing communities and applicators about the risks of DDT exposure, employing better application techniques and formulations, and studying the factors that determine direction.
- New vector control strategies should be developed and rigorously tested, considering regional differences in vectors, parasites, ecology, and culture. As with DDT, the effectiveness of novel vector control measures should be evaluated, as should any potential detrimental effects on the environment and population, including vulnerable subpopulations such as the malnourished and immunocompromised. Pyrethroids, for example, have been employed in place of DDT in interior residual spraying, although nothing is known about how they influence human health.

The most recent research on DDT exposure to human populations and its potential health effects supports the Stockholm Convention on Persistent Organic Pollutants, which emphasizes that DDT should be used with caution, only when necessary, and when there are no other effective, safe, and affordable alternatives locally available. The agreement requires each country that currently uses DDT to submit an implementation and management plan that limits DDT use to disease vector control and reduces reliance on DDT. Governments should get support to rely on other long-term malaria control approaches, methodologies, and strategies. Given the lack of data on populations potentially exposed to high levels of DDT, we implore the global community to monitor DDT exposure. They should also evaluate its potential health effects in malaria-endemic regions where DDT use has historically remained high, like the Pine River Superfund site.

Organophosphorus Pesticides

Glyphosate is by far the most widely used organophosphate pesticide. These chemicals were marketed as safer alternatives to organochlorines. Dimethoate, parathion, and malathion are well-known insecticides in this class. Many of these are associated with hormonal imbalances. This pesticide has been linked to genotoxicity, mitochondrial dysfunction, disruption of normal cellular metabolism of proteins, carbohydrates, and fats, changes in insulin secretion, neurological and endocrine system issues, and effects on cellular oxidative stress.

According to large-scale population studies, organophosphorus pesticide exposure has been linked to dementia, cardiovascular disease, adverse effects on the brain and male reproductive system, and an increased risk of non-lymphoma. Hodgkin's Furthermore, prenatal exposure to organophosphates has been linked to shorter gestational periods and neurological problems in children.

Glyphosate

Even more so now that glyphosate-tolerant GMO crops such as soy and maize have been introduced, the widespread use of glyphosate as an herbicide remains a source of contention among scientists. Concerns have been raised about the widespread use of glyphosate in the cultivation of genetically modified soybeans due to the potential synergistic estrogenic effects of glyphosate and the phytoestrogen "genistein," a common isoflavone found in soybeans and soybean products.

Glyphosate may affect human erythrocytes *in vitro*, the endocrine system and the development of skin cancer in experimental animals. Furthermore, the shikimate pathway, found in bacteria, plants, and human gut microorganisms, is thought to be severely disrupted. This disruption could impair the body's ability to obtain essential amino acids. The active ingredient is considered safer than regular glyphosate products when used alone. Exposure to glyphosate-containing herbicides, such as the popular "Roundup," has been shown *in vitro* to affect human skin cells, cause DNA damage in human cell lines, and alter the endocrine system in rat testicular cells. They can also hasten the death of testicular cells in laboratory animals. New research suggests that they can influence cellular transport and the cytoskeleton.

A recent study investigated the potential link between glyphosate, GMOs, and declining health in the United States. According to correlation research, glyphosate has been linked to various diseases and disorders, including cancer, hypertension, diabetes, strokes, autism, kidney failure, Parkinson's disease, and Alzheimer's disease. Glyphosate has also been linked to gluten intolerance, which has been linked to a lack of essential trace metals, difficulties during pregnancy, and an increased risk of developing non-Hodgkin's lymphoma cancer.

Carbamate Pesticides

Carbamate pesticides such as aldicarb, carbofuran, and ziram have been linked to an endocrine system disruption, possible reproductive disorders, effects on cellular metabolic processes, and mitochondrial function. According to *in vitro* research using hamster ovarian cells, carbamate pesticides can cause apoptosis and necrosis in human immune cells such as natural killer cells and T lymphocytes.

The carbaryl pesticide, a member of the carbamate pesticide class, can bind to the hepatic aryl hydrocarbon receptor, a transcription factor involved in the mechanism of dioxin toxicity. Carbamate pesticides have also been shown to impair neurobehavioral function, increase the risk of dementia, and aggravate non-Hodgkin lymphoma.

Other substantive chemical pesticides

Triazines, a class of chemical pesticides linked to endocrine disruption and reproductive toxicity, include atrazine, simazine, and ametryn. Further research revealed a possible statistical link between triazine herbicide exposure and the development of breast cancer. Atrazine, the most commonly used triazine herbicide, has been linked to dopaminergic effects, cytotoxicity, and oxidative stress. In experimental animals, atrazine exposure has been linked to reproductive toxicity and a delay in sexual maturation.

Synthetic pyrethroids fenvalerate, permethrin, and sumithrin are among the least harmful insecticides used in farming and public health today. However, evidence suggests that they can alter reproductive parameters and disrupt endocrine function in animals. A recent study found associations between multiple pyrethroid metabolites and DNA damage in human sperm, raising concerns about potential reproductive harm. Concerns have also been raised about these compounds' potential neurotoxicity during development.

The most commonly used insecticides are neonicotinoids such as imidacloprid, thiacloprid, and guadapyr, which are relatively new and were initially marketed as safe for use around non-target organisms. Their impact on bee populations is just one piece of evidence pointing to the contrary. There is some evidence that animals' endocrine and reproductive systems may also be affected. Another recent study discovered that neonicotinoids could increase aromatase expression, a critical development enzyme linked to breast cancer. More sustainable agricultural practices that protect human health are urgently required.

Chemicals are widely recognized as a threat to human health, wildlife populations, and ecosystems due to their widespread production and use in modern agricultural practices. Because agrochemicals pose serious risks to human and environmental health, immediate strategic action is required to phase them out and replace them with less hazardous alternatives. Furthermore, modern farming methods should pose less of a threat to both people and the environment. In terms of long-term viability, adopting farming practices that meet these standards outperforms precision agriculture.

However, Farmers must reduce reliance on agrochemicals by using them sparingly and only when necessary while also accounting for the spatial and temporal variability of all soil and crop factors within a given field. Natural phenomena such as yield, field, soil, and crop variability contribute to this uncertainty. Access to various sensors, GPS, and GIS can be beneficial. We strongly oppose the practice of precision agriculture, which led to the development of such tools. However, we believe that certain technological tools can reduce the risk of environmental and water pollution. In contrast, the farmers can amplify the financial benefits of reducing the use of chemical products (Menjivar et al, J, 2018).

To achieve this goal, It is unnecessary to replace human labor with a series of mechanized procedures. Instead, we believe that modern technological approaches to pesticide reduction will eventually result in less chemical use, if not total elimination, in various contexts, including urban green spaces. Mals, an Italian town near the Swiss and Austrian borders, is frequently considered an example of the rest of Europe. Because the Mals prohibit the use of pesticides, all of their crops are grown naturally. Pesticides were opposed by more than 70% of Mal's voters in a referendum held in 2014. This groundbreaking decision adheres to the principle of food sovereignty and highlights the importance of education and communication in raising public awareness and obtaining consent (discussed below).

For example, the international "Via Campesina" movement pioneered the democratic concept of food sovereignty, which has been a part of the sustainability movement for over two decades. Everyone has the right to develop their own national and local policies to end poverty, malnutrition, and hunger and protect cultural heritage and the natural environment; t was declared in 2007 in the African village of Nyéléni in Mali.

The negative economic, social, and environmental consequences of agriculture industrialization are beyond the control of locals. Industrialization and its attendant food surpluses, export-driven monocultures, low-wage labor demand, and other factors have not proven effective in combating hunger and malnutrition. However, the current agricultural model contributes significantly to food inequity, unsafe chemical use, inefficient use of resources such as land and water, and environmental devastation. Food sovereignty promotes social, economic, and environmental sustainability by protecting indigenous peoples and producing food for local markets. Control over one's food supply is increasingly regarded as a fundamental human right.

As the current agricultural model becomes the norm, chemical use has increased. Farmers and small producers in developing countries bear the brunt of industrial agriculture's consequences as multinational agrochemical corporations purchase large farms. To increase output and remain competitive in the market, it is frequently necessary to resort to harmful practices that endanger the natural world. Despite the importance of international environmental standards, food sovereignty does not require organic food production or limits on pesticide levels. Using agrochemicals is critical in giving customers a say in food production.

Discussion

Pest management is essential and dates back to antiquity when organic and inorganic substances were utilized as insecticides. Many chemical pesticides have been developed since then. Today's multinational agrochemical corporations—which essentially manage global food production—apply new compounds with pesticide properties and advance biotechnology, breaking from traditional agricultural practices. Furthermore, the foundation of current agricultural systems is widespread chemical pesticide use, which has been linked to negative impacts on human health, wildlife, and the environment.

Population growth, food security, the dangers of chemical pesticides to human health, pesticide resistance, environmental degradation, and climate change are all important challenges that contemporary agriculture must handle. In recent years, new concepts regarding agriculture and food production have developed. Climate-smart agriculture is a concept that seeks solutions in the emerging climate-change environment. Another key ongoing dispute between proponents and skeptics of genetically modified pesticide-resistant plants concerns both the safety and the impact of these plants on pesticide use. In-depth research is also required due to real-life chronic pesticide combination exposure, which may have additive or synergistic effects. The real complexity of the situation is exposed by the underlying scientific uncertainty, the exposure of vulnerable people, and the sheer number of possible combinations. Combining chemicals that are likely carcinogenic or have endocrine disruptive qualities may result in unknown harmful health effects. It is critical to consider chronic exposure to numerous chemicals when determining "acceptable" amounts of exposure to particular pesticides.

When the effects of chemical pesticides on human health and the environment are considered, it is clear that a new idea in agriculture is desperately needed. Because the European Common Agricultural Policy (CAP) predicts health, environmental, and economic benefits, this new concept must be based on a large reduction in chemical pesticides.

Adopting a comprehensive ban at the local level is a straightforward way to create pesticide-free regions in urban green spaces. Furthermore, new agricultural policies targeted at sustainable development and consumer health protection should be formed as alternatives to the current food production paradigm. Regardless of the difficulties associated with implementing a fresh notion, the associated parties must implement a new, safer, and cleaner farming paradigm (Menjivar et al, J, 2018).

3.4 BIOMAGNIFICATION AND BIOACCUMULATION OF CONTAMINANTS IN AQUATIC ANIMALS

3.4 1 Bioaccumulation

Bioaccumulation occurs when an organism absorbs a hazardous material (such as pesticides or chemical compounds) faster than it can excrete it. What exactly is the problem with bioaccumulation? When the body absorbs toxins faster than it can rid them, the danger of chronic poisoning rises. Even if the concentration of a harmful substance in the environment is modest, accumulation in the food chain can have fatal consequences for species. When the amount of toxin present in a certain environment is known, models can be used to analyze the possible accumulation in fish and mammals and predict the effect on the animals. It is good that scholars' predictions can be made, but that does not change the effect it will have on living organisms (Đolić et al, 2022).

Toxins, both natural and artificial, can accumulate in living things. Natural poisons include red tidesma, which are caused by algal blooms, and ciguatera poisoning (coral reef algae poison). It affects mussels and oysters when they filter feed on the hazardous plankton during red tides, and it builds up in coral fish when they eat the algae. As a result of this adaptation, several animal species use bioaccumulation as a kind of defense. When exposed to dangerous plants or animals, several species accumulate poisons in specific physical locations as protection. The tobacco hornworm, which chews tobacco plants and absorbs nicotine from them, is an example of this. Compounds not normally considered toxic may occasionally change when present in large quantities. The best example is the buildup of vitamin A in the livers of carnivores (meat eaters) such as polar bears. Polar bears will accumulate extremely high levels of vitamin A because they only eat other carnivores (seals). Polar bears are unaffected, but people who have consumed their liver have developed hypervitaminosis.

Natural toxins have far less of an influence on animals than the accumulation of artificial toxins. This can be seen in humpback dolphins in South Africa (*Sousa* sp.). Humpback dolphins are considered sensitive due to the number of chemicals that enter their systems. Humpback dolphins eat fish from rivers and reefs near the shore. Poisons from the fish in these habitats accumulate in dolphin blubber (fat). The biggest concern with this is the women's use of their fat reserves to generate milk for their calves. The mothers unintentionally poison their calves by feeding them poisoned milk, giving them a bad start (Đolić et al, 2022).

The physical-chemical properties of harmful molecules, environmental conditions, exposed organism features, and the organism's food chain are four variables that influence bioaccumulation. As a result of these characteristics, numerous bioaccumulation potentials may work together or against one another. Chemistry-Related Factors: The physical-chemical properties of the contaminated molecule have a significant impact on the bioaccumulation process. Chemical properties that boost their proclivity to bioaccumulation include the following:

(1) Lipophilicity or fat loving refers to the degree of a chemical's solubility in octanol, which is determined by the size of the octanol-water partition coefficient (kOW).

(2) The lack of polar functional groups results in low water solubility or hydrophobicity.

(3) Environmental permanence results from structural stability (years instead of days).

Finally, compounds that lack ionizable functional groups and have intermediate molecular weights and sizes (350 and less than 10 Angstroms, respectively) are more likely to bioaccumulate.

Elements of Environmental Importance

As previously stated, substantial levels of bioaccumulation are not frequently the result of compounds in the environment that match most of the conditions mentioned earlier. A major explanation for this decrease in bioaccumulation is low residue bioavailability. For bioaccumulation to occur, a chemical must come into touch with a bio-membrane and pass through the membrane to reach lipid-rich storage sites. The amount of chemical that comes into touch with an organism's absorbing membranes is determined by the percentage of chemical available for uptake and the chemical's ambient concentration in the bulk water phase (including particles) (Le., the bioavailable fraction). Typically, this accessible part corresponds to the portion of the chemical dissolved in water. Because particulate organic carbon in suspended and bed sediments shares some of the same chemical characteristics as lipids, lipophilic or bioaccumulative compounds have strong affinities for organic particle carbon. In an aquatic system, the bulk of a highly lipophilic contaminant's mass is often sorbed on particulate organic carbon rather than dissolved in the water. The desorptive release of a lipophilic residue from sediment organic carbon, which can occur very slowly, can considerably reduce the number of chemicals accessible for bioaccumulation. However, in locations where contamination inputs have declined, the principal source of trace levels of chemicals that undergo bioconcentration is the delayed release of contaminant-sorbed sediment.

The environmental stability or persistence of molecule substantially impacts its proclivity to bioaccumulate. Environmental degradation processes typically yield more hydrophilic or polar byproducts with lower bioaccumulation potentials on contaminating molecules than the parent chemicals. This result has some exceptions, notably when oxygen levels are low. However, the overall effect of these processes is to reduce organism exposure durations and parent compound concentrations, which reduces the amount of bioaccumulated residues that aquatic animals take up.

Factors Concerning the Organism's Lipophilic contaminants by breathing, food consumption, or sediment absorption. Most chemicals, including most PAHs, are assumed to be taken predominantly by bioconcentration by aquatic animals. Because fish breathe liters of water over their membranes every day, the primary point of contamination entry into an aquatic organism is often through the gills. This approach can assimilate anywhere from 20 to 90% of pollutant residues in vented water using a range of lipophilic compounds. Diet is more likely to be the predominant route of absorption when substances are persistent and have a high K content (Le., greater than IO5). This is especially true for top predators in a food chain, both aquatic and terrestrial. The quality of the substances consumed influences how well lipophilic chemicals are absorbed throughout the gut. If ingested materials are predominantly indigestible, such as the majority of organic carbon containing non-natural sediment, the probability of gastrointestinal uptake is decreased. Gut absorption efficiencies for several lipophilic substances from high-grade fish meals (such as animal or plant tissues) have been demonstrated to range between 50 and 85%. It should be noted that the fat content of the consumer organism affects an organism's final potential to accumulate a chemical but has little to no effect on chemical dietary and respiratory absorption rates (Roccaro, P., & Vagliasindi, F., 2014).

Bioaccumulation occurs when a chemical's absorption rate exceeds its excretion rate. Many lipophilic chemicals are passively depurated by aquatic organisms' gills. The significance of this form of elution appears to be mostly determined by the nonpolar compounds involved. In general, the removal rates of these compounds are inversely related to their KOW. In many animals, the cytochrome mixed-function monooxygenase (MFO) system and the aryl hydrocarbon hydroxylase system are both in charge of the biotransformation of a variety of lipophilic chemicals, including PAHs (especially mammals, birds, and aquatic vertebrates). Remember that biotransformation products are frequently more hydrophilic and remove much faster than their parent substances. The liver of fish, birds, and mammals contains most MFO activity, and bile is employed to remove the more hydrophilic metabolites.

On the other hand, some PAHs and alkanes can undergo transformations that produce more harmful intermediates (including carcinogenic) than the parent compounds. This can occur even if the MFO system adequately detoxifies and reduces the bioaccumulation of several contaminants. The basic trend described below illustrates how different species differ in their ability to clear accumulated PAH residues through all processes: mammals, fish, crustaceans, and bivalve mollusks. Because of their high bioaccumulation and relatively low clearance rates of PAH and other pollution residues, bivalves are frequently used as sentinels or biomonitoring species (Roccaro, P., & Vagliasindi, F., 2014).

There are several effects of bioaccumulation as discussed below;

Bioaccumulation in Human Health

- Polychlorinated biphenyls (PCBs) have been found in high amounts in bluefish and striped bass on rare occasions. However, bioaccumulated mercury levels are common in marine species such as swordfish, sharks, and tuna.
- Toxic pollutant bioaccumulation endangers human health.
- Humans ingest significant amounts of hazardous substances when they consume species at the top of the food chain.
- Consuming contaminated filter-feeding marine animals such as mussels, oysters, and coralfish can cause poisons to accumulate in people's bodies.

Ecosystem

- Animals can convert one pound of food consumed into body mass, increasing toxin concentrations at every level of the food chain by nearly a factor of ten.
- Normally held in fat, bio accumulators are released into the bloodstream when an animal burns body fat for energy, causing injury to key organs and systems.
- Nursing babies eat these bioaccumulators because they are secreted from breast tissue during milk production.
- If bio accumulators remove keystone species from an ecosystem, such as predatory animals that regulate prey numbers, many species may become extinct.

Bioaccumulators include PCBs, PAHs, heavy metals, certain pesticides, and cyanide.

Efforts are being made to reduce harmful chemical bioaccumulation. Legislation prohibiting the discharge of particular compounds in water contributes to reducing the number of dangerous substances in the environment that can accumulate in the food chain. Furthermore, bacteria are genetically modified to consume hazardous chemicals such as mercury. These bacteria can eliminate chemicals straight from the environment.

3.4.2 Biomagnification

The word "biomagnification," which stands for "Biological Magnification," refers to the increase in toxic or dangerous compounds in food systems. Dirty or polluted environments typically emit these compounds. Pesticides such as DDT, heavy metals such as mercury and arsenic, and polychlorinated biphenyls (PCBs) are among the toxins that are eaten by organisms through the food they eat or pollution in their surroundings.

This phenomenon is the outcome of a chain of events. The digestives' ability to hold non-metabolized lipophilic pollutants decreases as contaminated prey is digested in predators' stomachs, resulting in a net transfer of these substances to the predator's lipid-rich tissues. Suppose the predator continues to ingest a variety of prey. In that case, the rates of uptake by the diet can then exceed the rates of elimination, resulting in contaminant concentrations higher than those seen in the predator's fatty tissues at equilibrium.

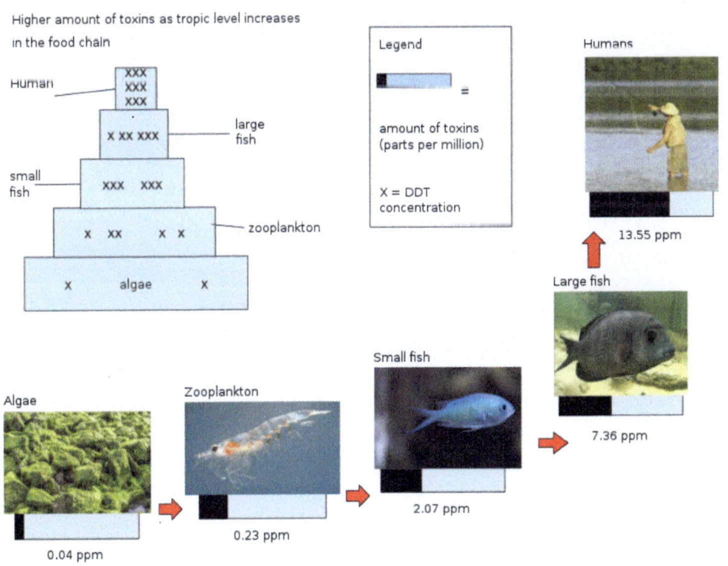

Figure 3.1. Biomagnification of pesticides (DDT) at different trophic levels in a food chain.

Source: By ***Øystein*** *Paulsen - This file was derived from: Meganyctiphanes norvegica2.jpg: Fly fisher in water fishing.jpg: & [1] & [2] & [3], CC BY-SA 3.0,*

The remaining concentrations may increase even more if a predator later devours this animal at a higher trophic level. When the predators are fish-eating birds and animals with high consumption rates of contaminated fatty prey and limited removal pathways, biomagnification can result in residue concentrations that are 100-fold higher than the equilibrium values.

Significant concentrations of these elements can be found in various industrial and home compounds. The organism's cells then begin to amass the hazardous chemicals. Toxins gradually build in the higher food chain due to ingestion by higher food chain organisms of animals below their trophic levels. Because this is a recurring process in the ecosystem and throughout the food chain, higher species will accumulate most of the toxins.

The following are some of the main causes of biomagnification

- Release of toxic substances in the environment

  Toxins and harmful substances accumulate in the environment due to toxic chemicals and pollutants released into habitats such as the oceans, air, and land. When these dangerous compounds and pollutants are introduced into various ecosystems, their concentration appears to be very low; yet, they accumulate over time and are eventually absorbed by lower organisms in the food chain, such as fish, earthworms, and plants. When higher organisms consume lower species, biomagnification occurs, and the process moves up the trophic levels.

- Agriculture

  Pesticides, fungicides, herbicides, and chemical fertilizers are among the many agricultural chemicals that infiltrate soils, rivers, lakes, and the ocean through surface stormwater runoff. Heavy metal traces such as arsenic, cadmium, mercury, copper, and lead can be detected in common agricultural inputs such as pesticides, industrial byproduct wastes, some fertilizers, and specific agrochemical products. When these substances are taken indirectly, they harm human health and the health of aquatic species such as fish because they accumulate in bodily tissues.

- Organic contaminants

  Manures and biosolids often contain nitrogen, carbon, phosphorus, and other nutrients. Because they have been industrially processed, they may also contain pollutants such as pharmaceuticals and personal care products (PPCPs). These compounds, found in the bodies of humans and animals, are regarded to be detrimental to wildlife, animals, and humans.

- Commercial manufacturing activities and pollution

Industries' indirect or direct industrial operations produce harmful and poisonous substances in the soil, rivers, lakes, and seas. Industrial operations contribute to environmental contamination in various ways, including releasing harmful chemicals into the atmosphere, which enter the food chain. Exhaust gases from cars and enterprises that make and refine oil can dissolve in the rain and cause acid rain, contributing to air pollution.

The soil and water bodies absorb the chemicals in acid rain. They travel up the food chain after being consumed by primary consumers, causing biomagnification.

- Mining activities in the ocean

Efforts to gather minerals and metal ores such as zinc, cobalt, silver, aluminum, and gold from the deep sea are harming the seas and coastal regions due to the many sulfide and selenium deposits formed by mining processes. The accumulation of heavy metals at various trophic levels is caused by an increase in hazardous levels absorbed by marine life, which is then consumed by organisms in higher trophic levels.

- Plastic pollution

The disposal of plastic rubbish near or in bodies of water may be harmful to organisms higher up the food chain and aquatic organisms. According to a study, one of the main toxins introduced into water bodies is Bisphenol A, which is in plastic.

Figure 3.2. Polluted sandy beach.

Source: Guffogg JA, Blades SM, Soto-Berelov M, Bellman CJ, Skidmore AK, Jones SD. Quantifying Marine Plastic Debris in a Beach Environment Using Spectral Analysis. *Remote Sensing*. 2021; 13(22):4548. https://doi.org/10.3390/rs13224548

Other people who consume the fish consume the tiny pieces of shattered plastic inside the fish's body.

Below are some of the effects of biomagnification

Impact on human health

Mercury, cadmium, lead, cobalt, chromium, and other chemical toxins make humans more susceptible to cancer, liver and kidney failure, respiratory disorders, congenital disabilities in pregnant women, brain damage, and heart problems. For example, eating seafood tainted with polycyclic aromatic hydrocarbons (PAHs) and mercury has been related to ailments such as hepatitis and cancer.

Implications on the growth and reproduction of marine organisms

The various harmful compounds and components accumulate in the critical organs of the various aquatic species, affecting their ability to reproduce and develop. Seabirds, for example, may choose to break their eggs rather than incubate them since their eggs have thinner shells than typical. Heavy metals such as mercury and selenium affect fish reproduction because they destroy their reproductive organs. Furthermore, PCBs (polychlorinated biphenyls), which are much more widespread in aquatic systems, biomagnify, inhibit reproduction, and are hazardous.

Destruction of the coral reefs

Cyanide, used in gold extraction and fishing, devastates coral reefs. Many water organisms use the reefs for spawning, eating, and dwelling. When aquatic animals are destroyed, their ability to survive is severely limited.

Food chain disruption

Many marine organisms rely on the natural food chain for survival. When poisons and other compounds are distributed into soils, rivers, lakes, or seas and swallowed by various species, the interdependent relationships within the food chain are disrupted.

It happens when little animals or plants consume or absorb toxic compounds, which are consumed by larger organisms and affect the entire natural food chain. Humans and higher-ranking animals in the food chain may be more prone to sickness, reproductive issues, and even death if they consume animals and plants poisoned by metals such as mercury, copper, chromium, selenium, and cobalt.

Poisonous chemicals such as heavy metals, pesticides, or polychlorinated biphenyls (PCBs) can travel up the food chain by infiltrating soil or water systems, where they are eaten by aquatic organisms or plants, which are then swallowed by smaller animals, humans, and huge birds. Because these substances are slowly ejected or metabolized/broken down, their concentration in organisms steadily increases as they go up the food chain.

The process of biomagnification includes the following:

- *Pollutants and hazardous chemicals released into the environment.* Toxic compounds and contaminants are initially released into the environment, eventually making their way into soils, rivers, lakes, and the ocean. The level of these hazardous chemicals appears to be rather modest when emitted in diverse circumstances.
- *Phytoplankton.* Phytoplankton is little plants that float in the water and absorb contaminants. Poisons that have been absorbed do not leave the body's tissues via excretion or disintegration. The poisons accumulate over time to high concentrations of up to 200 parts per trillion, resulting in a fourfold increase in toxicity.
- *Zooplankton.* Zooplankton is little marine animals that float in the water. They consume the toxin by eating phytoplankton. Poisons remain immobilized in the organism's tissue unless removed or destroyed. The concentration of poison increases over time, eventually reaching two parts per billion, or nearly ten times the starting concentration.
- *Small fish consume the zooplankton.* When the little fish devour zooplankton, the toxins are absorbed into their fatty tissues. As a result, accumulating occurs, raising concentrations by tenfold to around 20 parts per billion.
- *Large fish eat smaller fish.* Toxins that accumulate in the fatty tissues of smaller fish are ingested by large fish when they graze on them for food. Concentrations might reach 80 to 100 parts per billion or higher as concentrations rise. The dangerous levels have climbed four to five times.

- *The top of the chain consumes the fish.* When large fish are consumed, top-of-the-food-chain organisms such as dolphins, seabirds, and humans eventually accumulate toxins in their tissues, such as the liver. In this region, concentrations reach 10,000 to 15,000 parts per billion. Because they interfere with the healthy operation of critical organs, the effects impair the animal's capacity to reproduce and make it more prone to infections.

Below are biological magnification and bioaccumulation similarities

- Toxic chemical accumulation can occur at various levels of a food chain via two distinct processes: bioaccumulation and biological amplification.
- They describe how dangerous compounds go up and down the food chain's trophic levels. These hazardous chemicals usually accumulate in the tissues of living beings.
- The variables that determine the concentration of dangerous compounds at a certain food chain level are persistence, food chain energetics, and excretion rate.

The following are the differences between biological accumulation and biological magnification

Definition: Bioaccumulation refers to the continuous accumulation of contaminants such as pesticides or other chemicals in living things. Biological multiplication, on the other hand, depicts the buildup of harmful material in the tissues of tolerant species at progressively higher levels of a food chain.

Importance: Biological magnification, as opposed to bioaccumulation, refers to the process through which dangerous substances accumulate in the bodies of living things as they move up the food chain.

Correspondence: Bioaccumulation occurs when organisms consume or absorb hazardous compounds via the skin. Regarding biological magnification, hazardous chemicals start at minute levels at the bottom of the food chain and gradually increase in quantity as the food chain climbs.

Importance: Furthermore, biological magnification generates higher concentrations of dangerous compounds at higher trophic levels of a food chain. In contrast, bioaccumulation causes chronic poisoning even when the toxic component concentration in the surrounding environment is not too high. In conclusion, the buildup of hazardous compounds in living things is known as bioaccumulation. In essence, these chemicals love fat and build up in the fat tissue. At a certain dosage, bioaccumulation causes these compounds to become chronically toxic, resulting in substance-specific symptoms. In contrast, biological amplification is the accumulation of poisonous chemicals along a food chain's trophic levels. As a result, the poisonous material is present in lower concentrations down the food chain's trophic levels than at the highest trophic level. Therefore, the major distinction between biological amplification and bioaccumulation is their relevance.

Chapter

4

CROP CONTAMINANT TYPES, DETECTION, MONITORING

Contents

4.1 Contamination By Heavy Metals..68
4.2 Food Contamination in Agricultural Soil 72
4.3 Application of Manures and Bio-Solids 74
4.4 Detection..75
4.5 Monitoring..76

4.1 CONTAMINATION BY HEAVY METALS

Stormwater is considered a very reliable resource in various parts of the world more so in areas where water is not an easily found resource. In most areas, the stormwater is harvested and used in irrigation. It is mostly used in the sustainable production of food crops. In as much as using stormwater is very beneficial, there is a challenge presented by its use, especially in areas with high levels of pollution.

Environmentalists have expressed their concern with regard to the use of stormwater for various uses as it has been found to have a strong concentration and toxicity profile. Stormwater contains very important metals needed by plants. However, they are found in extremely large amounts. Some other metals found in stormwater are harmful. They include lead, mercury and arsenic. Metal uptake in plants from other water sources may face some challenges, this is not the case with storm water. It is known to pose some inherently variable characteristics that make metal uptake much faster and easier in plants (de la Guardia, M. 2021).

Figure 4.1. Immobilization of heavy metals: an example.

Source: Correia AAS, Matos MPSR, Gomes AR, Rasteiro MG. Immobilization of Heavy Metals in Contaminated Soils—Performance Assessment in Conditions Similar to a Real Scenario. Applied Sciences. 2020; 10(22):7950. https://doi.org/10.3390/app10227950

Among the emerging issues under food security and environmental health is contamination of heavy metals which poses a great threat to human beings and the environment at large. There are many risks associated with heavy metal contamination. Contamination of food mostly occurs in the growth process: as fruits and vegetables are planted and harvested. The environment is involved as food comes from our environment meaning that if the soils and water are contaminated, the same is the case for food.

One of the heavy metals contaminating food is lead, it is termed inconsequential to organic and metabolic capacities. However, it is important to note that certain metals are very useful to plants. They include zinc and iron among others. They are involved in metabolic cycles and catalyzing reactions and to some extent, they are related to biota metabolism. Certain heavy metals are useful and in some concentrations pose a threat with regard to food contamination. This is in the case of Nickel. Though it plays a vital role in plants, it is hazardous to human health.

In food production, the largest food contamination by heavy metals is soil. This is attributed to the fact that soil is very vulnerable to contamination by heavy metals. There are various sources of soil contamination that includes;

- Coal mine shafts
- Non-point sources such as spillovers from farms
- Nuclear energy stations and chloro-soluble base synthetic ventures

Soil contamination leads to food contamination by affecting microbial cycles which in turn affect soil biota. Studies done on plants exposed to modern or refining environments show that such plants tend to carry and cause the bioaccumulation of different kinds of heavy metals.

The occurrence of heavy metals in the environment is through natural sources and anthropogenic sources. These heavy metals find their way to food through ground water or soil. They eventually find their way to food webs and cause bioaccumulation. As heavy metals are toxic in small concentrations and are highly reactive, they pose a threat to biota. Research shows that the effects of heavy metals on biota differ depending on the type of heavy metal contaminating the soil and the contamination is very distinct. The various implications of the negative use of the environment have contributed to the natural contamination of soil. The high contamination of soil is attributed to various activities that include improper removal of rubbish, disposing of organic wastes in water bodies, improper use and management of pesticides and manures, and increased growth of metal enterprises and horticultural activities. (Subhashree, S., & Kumar, P. S. 2022). There are various factors that contribute to the increased contamination of food by heavy metals. They include climate. Climate as an environmental factor covers aspects of the hydrosphere, atmosphere and land. For this reason, the earth is made of four main components that include water, land, biosphere and air. All of these components work together to allow the sustenance of plants and human beings. Climate as an aspect of the environment carries its own natural toxins. When not properly balanced, these toxins can get to exceedingly high levels. High levels of toxicity are attributed to the process of industrialization that has grown throughout the years. It has greatly contributed to the exploitation of natural resources. Environmentalists have expressed their concern for ecological contamination. Industrialization has led to the poisoning of climate as an environmental factor through the release of radioactive substances, inorganic particles, nano-based materials, vaporous poisons and natural toxins. There are various ways of food contamination through heavy metals. They include contamination through irrigation water and soil among others (de la Guardia, M. 2021).

Food contamination through irrigation water

Food contamination by irrigation water means that water used to irrigate plants in agricultural lands are loaded with the heavy metals. There are various ways in which irrigation water can be contaminated. It includes the use of wastewater in irrigating plants. Ground water used in irrigating crops can also be contaminated. Ground water contamination occurs due to substrate erosion. Substrate erosion causes the release of these metals into water which eventually finds its way to plant roots. Over the years, the increased rate of human activities has caused an increase in substrate erosion. Wastewater more so from factories contributes to the heaviest infestation in edible parts of crops. Contamination through irrigation water can also happen through contact between wastewater and soils. As wastewater is deposited in soil, the heavy metals accumulate over time leaving the soil heavily contaminated. This mostly occurs in densely populated areas as wastewater is used in growing crops.

The use of contaminated water in irrigation can also be through the use of stormwater in irrigation. Stormwater is considered a very reliable resource in various parts of the world more so in areas where water is not an easily found resource. In most areas, the stormwater if harvested and used in garden irrigation. Stormwater is mostly used in the sustainable production of food crops. In as much as using stormwater is very beneficial, there is a challenge presented by its use more so in areas with high levels of pollution. Environmentalists have expressed their concern with regard to the use of stormwater for various uses as it was found to have a strong concentration and toxicity profile. Stormwater contains very important metals needed by plants. However, they are found in extremely large amounts. Some other metals found in stormwater are harmful. They include lead, mercury and arsenic. Metal uptake in plants from other water sources may pose some challenges, this is not the case with stormwater. It is known to pose some inherently variable characteristics that make metal uptake much faster and easier in plants.

Wastewater is gotten from industries including the packing, textile, garment, dyeing and pharmaceutical industries. Most of these processes make use of water in large amounts. The problem, however, is the fact that most of these industries do not treat their wastewater before releasing it to the environment. The untreated wastewater is usually deposited in nearby rivers or water canals. The untreated wastewater eventually finds its way to streams used in irrigation. This has been noted in various countries including Bangladesh. Bangladesh has two major districts known for their involvement in industrial processes. Their industries are known to release wastewater to river canals. For this reason, their soils and surface waters have been found to be rich in various heavy metals including lead, copper and zinc. The farmers from Bangladesh then use the contaminated water to irrigate their crops. This leads to the heavy metal contamination of their vegetable produce. Their vegetable produce has been found to have high amounts of Cadmium, Copper, Iron and Zinc among others.

Heavy metal contamination of food stuffs also occurs in animal and milk products. It has also increased over the years such that it has become a matter of great concern. The growing concern could also be attributed to the fact that a significantly large number of animals have died due to heavy metal contamination. The challenge of heavy metal contamination in animals is that the heavy metals accumulate and do not subside as they are non-biodegradable.

Over the years, several animal products such as meat and milk have been contaminated. Heavy metals find their way into the bodies of animals such as cattle, sheep, goats, hens and pigs by eating contaminated grass, water, plants and soil. Eating contaminated grass and water is one of the main ways through which heavy metals find their way to livestock.

Various heavy metals are released in livestock products. For instance, lead is not easily excreted through milk. Lead is mostly found in milk when the animal's blood is contaminated by lead. For milk from livestock to have lead concentrations, the animals must have been fed on concentrates and grass having high levels of lead. Generally, livestock whose bodies have high concentrations of heavy metals are usually found in contaminated environments. Feeds having mercury and cadmium when fed to chicken leads to heavy metal contamination.

In livestock production, the quality of the animal product is dependent on the health of an animal and its age. Studies show that high levels of heavy metal contamination are high in animals that are old, young and undernourished animals that are raised in stressed conditions. Contamination of heavy metals in animals varies depending on the type of animal. For instance, goats are more likely to suffer from lead contamination while cows are likely to suffer from arsenic contamination which leads to arsenic contamination of milk. Poultry products are most likely to pose high levels of arsenic.

To prevent food contamination by heavy metals in livestock, agricultural ministries place laws that required livestock feeds to be investigated for their components before it is fed to animals. The challenge however is that most of the testing sites are not accessible to some farmers. Animals are also exposed to heavy metals when they drink waters that have been contaminated. This has allowed the exportation and importation of contaminated livestock products. Medication administered to livestock is also one of the routes of heavy metal contamination in animals. Over the years, there are some drugs that have been banned for use among animals as they contain some heavy metals such and lead and mercury and when not properly administered to animals could lead to heavy metal accumulation over a given period of time (de la Guardia, M. 2021).

4.2 FOOD CONTAMINATION IN AGRICULTURAL SOIL

Food contamination can also occur through the type of soil used in planting the crops. In this case, soils that produce contaminated food crops are already contaminated by heavy metals. There are various sources of heavy metal contamination in soil that include;

- Petrochemical spillage
- Animal excrement
- Mine tailings
- Sewage slop
- Leaded paints and gas

- Coal burning buildups and composts

Some of the heavy metal pollutants in soil include mercury, lead, nickel, chromium, copper and arsenic. The challenge with soil contamination is that the heavy metals do not degrade over time as they do not go through compound or microbial degradation. They instead accumulate over time. The presence of heavy metals in soil affects the bioavailability and structure of soil. Most of these heavy metals affect the biodegradation of regular soil pollutants. It also affects the organic framework of animals in soil (Subhashree, S., & Kumar, P. S. 2022).

Heavy metal contamination of soil has greatly contributed to the decline in food quality rendering land unusable in agricultural practices. Heavy metals lead to the generation of phytotoxicity. There are various reasons why heavy metals eventually become contaminants in soil. It includes the soils being transferred from mines into various areas where they could be exposed to contaminants, presence of chemical components where when soil comes into contact with it could render the soil more bioavailable and there can be an easy transfer of contaminants in soil. Studies show that heavy metals such as lead and zinc in soil from anthropogenic sources are usually more mobile. This means that such heavy metals are more bioavailable than pedogenic. Some contaminated soils are naturally occurring as by nature they are metal-bearing solids. Some soils from such contaminated sites can be moved to other areas leading to further contamination of soils.

The means by which agricultural soils are contaminated are as follows:

- Fertilizers used in agricultural processes

One of the major ways in which human activities have affected soil health is through the use of fertilizers. Fertilizers are one of the tools used by farmers in boosting plant growth. Fertilizers are fitted with various nutrients that facilitate the growth and allow the lifecycle of animals in plants. Fertilizers are made up of both micronutrients and macronutrients and they all play vital roles in the growth and germination of plants. In as much as heavy metals lead to the contamination of plants, some of them are needed to ensure soil health. There are certain soils that are deficient in the heavy metals needed to ensure the proper growth of plants. For such soils, it is very necessary that farmers use fertilizers to instill the nutrients in soil. In most cases, intensive farming practices involve heavy use of fertilizers which eventually leads to soil contamination. Intensive use of fertilizers is usually done to add potassium, sodium and nitrogen in the soil. In most cases, when these elements are used in soil they are usually accompanied with trace amounts of certain heavy metals. When fertilizers are accompanied with trace elements and are used continually, the concentrations of the heavy metals increase over time. When such soils are used in agricultural practices, these soils lead to the contamination of plants leading to the supply of contaminated foodstuffs across the world. (de la Guardia, M. 2021).

Use of pesticides in agricultural lands

There are different kinds of pesticides used in agriculture. Most of these pesticides have certain amounts of heavy metals in various concentrations. Pesticides include fungicides and insecticides. Most of the pesticides are manufactured in the United Kingdom and are made using various heavy metals such as zinc, lead, manganese and mercury. Copper oxychloride and copper sulfate are some examples of pesticides made using copper. There are various fruit orchards known to make use of lead arsenate in dealing with parasitic insects. Pesticides made using arsenic are mostly used in controlling pests such as ticks. They are mostly used in Australia and New Zealand. In these countries, copper and arsenic are used to make formulations used in preserving timbers. However, when using these pesticides in most cases they are not properly used and lead to the contamination of soil. In a comparison between fertilizers and pesticides, the use of pesticides is more restricted to the area of application.

4.3 APPLICATION OF MANURES AND BIO-SOLIDS

The term used to collectively refer to municipal sewage sludge, compost and livestock manures is bio-solids. The continuous application of bio-solids on soil leads to the continuous increase in heavy metal concentration in soil. This leads to high levels of zinc, nickel, mercury and arsenic among others. Manures used in soil are usually from certain animals. Wastes from cattle, pigs and poultry are usually gathered together and used in agricultural lands in the form of slurries and solids. There is a common perception that manures are a good kind of fertilizer. However, they may contain certain amounts of heavy metals. For instance, wastes obtained from pig farms are very rich in zinc and copper as they are usually fed on diets rich in copper and zinc. Therefore, when such kind of wastes are used in farms, it leads to the accumulation of heavy metals in soil to an extent that soil becomes heavily contaminated with such metals (Subhashree, S., & Kumar, P. S. 2022). By nature, bio-solids are by nature organic solid products primarily made of wastewater treatment processes. There are a number of countries known to use bio-solids generated in urban areas. In some areas, bio-solids are majorly known as sewage sludge. As bio-solids are generated in large amounts in urban areas, very large amounts are deposited in agricultural lands as a means of clearing them out. Most of these countries use bio-solids as fertilizers.

Innovators are looking for ways of venturing into the use of bio-solids in composting alongside other organic materials including agricultural wastes. If this idea is to go through then there is a high probability that heavy metal contamination would increase to even greater concentrations such that foods produced from such soils would not be fit for human consumption. This means that heavy use of bio-solids poses a great threat to agricultural production. Among the heavy metals found in bio-solids include zinc, copper, nickel and lead. The amounts of these in the soil are governed by the type of industrial work done in the area. When some of these metals are added to soil, some of them may be leached into the soil through the soil profile. In some severe cases, the heavy metals may find their way into ground water thereby contaminating it. Studies done in New Zealand have proved that soils that had received bio-solid application had higher levels of heavy metal concentrations.

Continuous use of manures and bio-solids for a long term may eventually lead to damaged soil profiles and plant produce with high levels of heavy metal contamination. For this reason, manures used in farms should be checked to ascertain that they do not contain any amount of heavy metals (de la Guardia, M. 2021).

4.4 DETECTION

The most commonly occurring heavy metals found in food include lead, mercury, cadmium and arsenic among others. With detection of heavy metals in food, it is important to note that accumulation occurs gradually. This means that the detection may not be easily notable at the start. Plants present different symptoms of contamination. There is also a variation in the symptoms depending on the type of heavy metal found in the plant. If the source of contamination is not known, then more tests will have to be done to detect the presence of heavy metals in plants. Most of these heavy metals are very useful in the bodies of human beings. However, they should not be taken in very large amounts. Depending on the diet there may be different concentrations of heavy metals in the foods of human beings. As the exact knowledge of the duration of exposure is not well known, researchers tend to conduct regular high-sensitivity testing to determine the presence of these metals in food. This means that there is a higher frequency of testing to facilitate the detection of these metals. There are various metals involved in ensuring food safety among people. They include the Food Safety and Standards Authority of India, The European Commission, the World Health Organization, China Food and Drug Administration and China Food and Drug Administration. They have been tasked with the duty of regulating the consumption of heavy metals through consumer products and animal feeds.

There is a standard level of these heavy metals that is allowable by these boards. Certain densities are considered non-hazardous in different kinds of foods. However, over the years it has been noted that short-term exposures have increased over the years. This means that there are increased health risks. The release of unmonitored food substances into the market has allowed contaminated food substances into people's homes. Detection of heavy metals in foods mostly involves samples of foods being subjected to a series of tests some of which may require very expensive equipment and various chemical reactions being made. They are able to determine the right amounts of such foods that can be consumed by human beings.

More technological advancements have been made so that the detection of heavy metals can be achieved. Most of these technologies are analytical in nature. Further advancements have been useful such that it has allowed improved availability, sensitivity and efficiency in the testing of heavy metals in food. The various technologies allow a variety. Some of the analytical technologies make use of inductively coupled plasma-optical emission spectroscopy or atomic absorption spectrophotometry. The development of other technologies has allowed the tools to be highly sensitive in measuring the presence of heavy metals. People handling such kinds of tools need to be highly knowledgeable in management and use of such tools (de la Guardia, M. 2021).

4.5 MONITORING

Monitoring of heavy metals in food products involves continuous monitoring of levels of heavy metals in foods. This means that tests should be done at a given frequency. This will allow easy detection of heavy metals and any spice in its concentration would be easily noted.

Unwanted sediments and chemicals

Usually, contamination of food products by unwanted sediments and chemicals tends to go together. Contamination by unwanted sediments tends to occur in farms. Most of these sediments are found on the surface of these crops. They are transported to agricultural lands through wind or agricultural water used in irrigation. Wind carries away sediments from various places to agriculture grounds. They finally settle on the leaves of plants or the surface of fruits. When harvested, these crops are transported with the sediments on them and if not properly cleaned the sediments remain on food substances.

Unwanted sediments can also be found on food products during the production and packaging process. This happens when the conditions in which food contains the unwanted sediments. The location of food managing companies also determines the presence of unwanted sediments. If the area is located in windy and dusty areas, then there is a higher likelihood of it being contaminated by unwanted sediment. This kind of contamination can also occur in restaurants as people indulge in their food or when making the various dishes.

Chemical contamination is also very common. It is very common that it has become a global food safety problem. Chemicals can be found in various parts of the environment. Some of the chemicals are very toxic while others do not pose much risk to human health. Chemicals that contaminate foods can either be organic or inorganic in nature. In most cases, these chemicals find their way into food through environmental contaminants. The environment is one of the sources of chemical contamination. For instance, food is grown in soils and requires water for their germination. The presence of heavy metals such as copper, mercury and lead in the environment contributes to food contamination. Chemical contamination can also occur through the use of pesticides in the agricultural process or through the different drugs administered to livestock. Chemical contamination can also occur in the packaging process through the leaching of the chemicals from the materials used to pack them. Effects of chemical contamination may vary and range from acute to chronic.

There are various levels of food contamination determined by the nature of contamination. The nature of contamination may be due to various factors including contaminants from food packages, different exposures to pesticides and the manner in which plants take up the chemicals from the environment. The level of chemical contamination in animals is greatly determined by their dietary make-up. The toxic effects of chemicals in the body are also affected by the life stage of the individual. Chemical contamination of food stuffs can also occur through the presence of some heavy metals. Contamination of such a kind is really common in urban gardens and farms. Chemical contamination can also take the form of water contamination. Water is considered part of food. Chemical contamination of water can occur through spraying of pesticides near water sources or water bodies. It can also occur during the purification process. When consumed, such water finds its way into the body of animals and human beings. Chemical contamination can also occur in rainwater as is the case of Xenoestrogenic compounds found in water. Chemical contamination of water leads to the manipulation of marine and soil biota and has eventually played a role in tampering with the safety of sea food. This means that sea food will be rendered unsuitable for consumption by human beings.

Like other forms of contamination, there is no properly defined safe dose level of chemicals. As mentioned, certain chemicals are acceptable and therefore there is a given standard for the most appropriate amount of chemicals that can be used. For instance, when drinking water with chemicals, the presence of heavy metals such as lead and arsenic is difficult as it could lead to skin cancer. Long-term exposure to such kinds of water may eventually lead to the occurrence of such cancers. There are various channels of chemical contamination in food stuffs. To better understand how water and food are contaminated, it is very important to know the route in which they are eventually consumed. Some of the factors facilitating the presence of chemicals in water and soil include human activities such as mining. Activities such as mining allow the exposure of various soil substances including mercury and arsenic. Once released into the environment, they get into contact with various environmental elements including soil and water. Some of the chemicals that contaminate soil and water are as follows:

Metals

The occurrence of metals and their various metalloids in the environment are generated from various sources. Artisanal golf mining is known to contribute to the presence of lead and mercury in soil. This is well noted in areas such as China, Shaanxi and Tongguan. The mining activities have led to heavy exposure to such kinds of heavy metals such that their locally produced food produce have very high levels of heavy metals that exceed levels set by their governments. Morocco is known for its iron mines which have led to the exposure of cadmium and lead that has been highly found in livestock produce in relatively large amounts. Spain also has a few mines that have contributed to sheep product contamination due to the exposure of lead. Plant contamination by metals heavily occurs in areas that heavily practice industrial activities. Each country deals with different levels of toxicity from various metals. Crops in Romania have exceedingly high amounts of zinc, copper and lead among other metals.

Chemical contamination through metals also occurs in vegetables and fruits. This has been noted in China that high levels of cadmium are found in navel oranges due to its presence in high amounts in soil. Due to the various chemicals found in China soils, high levels of contaminated metals are found in edible seeds. In as much as these seeds are met to improve human health, they have been found to have high levels of copper that would affect human health. Some of these chemicals are found in water thereby affecting food products gotten from the sea such as fish. In countries such as Turkey, samples from fish showed that they contained high amounts of manganese, iron and copper. Usually, these metals are found in saline water. Mercury from industries found its way to Asian waters contaminating their turtles with metal. China has reported several cases of metal contamination in its drinking water. In some other parts of the world, metals such as nickel have been found in drinking water. (de la Guardia, M. 2021).

Contamination through polycyclic aromatic hydrocarbons

When organic matter is subjected to pyrolysis or incomplete combustion in industries, it leads to the formation of polycyclic aromatic hydrocarbons. Food contamination by these hydrocarbons is mostly through exposure to home cooking, polluted environments and industries. By nature, the polycyclic aromatic hydrocarbons are carcinogenic and genotoxic. Contamination may occur through the occurrence of accidents on water bodies that results in oil spills. This means that such kinds of contaminations usually affect sea animals. These hydrocarbons are mostly found in petroleum-based products. They are also found in various chemical dispersants used in the mitigation of oil effects on the ocean. For this reason, good samples are obtained from fish before they are sent to markets to ensure that they are safe for human consumption.

Contamination through industrial chemicals

Some of the chemicals used and produced in industries find their way to food stuffs leading to contamination. Some of these chemicals are organic chemicals. They include various organic pollutants. Some of the chemicals produced in industries include herbicides, fungicides and insecticides. Some of the chemicals are produced as by-products of the various reactions that occur in industries. They include dibenzofurans and dibenzodioxins while HCBs and PCBs are industrial chemicals. Their continuous release in the environment has allowed for continual accumulation of these chemicals in the environment. They are distributed to the various parts of the environment including oceans and air. They also infiltrate the food chain and do not leave. They circulate the food chain. There are various effects of the presence of these chemicals on all the affected animals and plants.

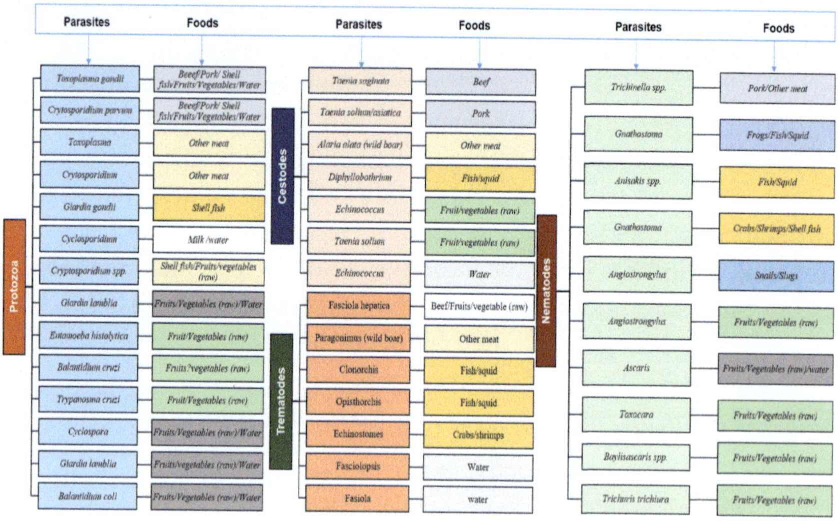

Figure 4.2. Parasites in different foods.

Source: Lim et al. (2017). The Sources of Chemical Contaminants in Food and Their Health Implications. Retrieved from Frontiers: https://www.frontiersin.org/articles/10.3389/fphar.2017.00830/full

Produced industrial chemicals such as Chlorpyrifos when consumed is known to cause various neurological and affect vision. This chemical is found in various dietary supplements as it is a well-known pesticide. It accounts for about 13% of daily exposure to chemicals. It is an example of an organophosphate. There are organo-chlorine pesticides commonly used in agriculture in controlling diseases transmitted by certain vectors. These pesticides are very popular but due to their harmful effects on humans and animal health, their use has been reduced. Other industrial chemicals such as deltamethrin and permethrin classified as pyrethroids are often used in conducting aircraft disinfection and controlling insect spreading diseases. In as much as pyrethroids are safe for use, their exposure to foods leads to food contamination though there are possibilities of reducing some of their toxicity.

Industrial chemicals include the release of PBCs that have extensive uses in various industries. They are used as paint additives in transformers. It is very easy for PCBs to find their way into human food more so livestock products. Contamination usually occurs in the form of residues. The danger with PCBs as chemicals is that they are readily absorbed by the body. It is often lined with the occurrence of Chloracne when taken in large amounts.

Natural chemical sources lead to the emission of polychlorinated and polybrominated compounds. These compounds are used in making dielectrics used in various electrical devices or as flame retardants. These compounds have found their way into human food. In China, its presence has been noted in human breast milk. These chemicals are usually exposed to human beings through their diets. Their release from industries into wastewater has made consumption of sea food the main pathway of human and livestock contamination. Some chemicals find their way to food through the various activities undertaken daily such as baking and heating. They include Acrylamide. They are usually released into the environment in the form of wastewater or when they are used as a chemical grouting agent. This chemical is very dangerous as it could lead to chemical contamination.

The final type of chemicals manufactured in industries are the PFCs. These chemicals are synthetic in nature. They are widely used in industries due to their friction-resistant properties. They are released to the environment in the form of wastewater which is often at times are used in agricultural lands. When ingested, some of the effects include problems in development and disruptions in the immune system. Their use however is linked to the production of polyfluoroalkyl and perfluoroalkyl substances. These substances are very toxic in nature. They are also released to the environment mostly as wastewater and tend to contaminate food products gotten from the sea. Human exposure to such chemicals is mostly through drinking water and contaminated sea food. However, there are minimal risks of heavy contamination (de la Guardia, M. 2021).

Contamination through pharmaceutical products

Pharmaceutical products are those that are manufactured in industries and are directly given to human beings, animals and plants. Most pharmaceutical products are made of certain chemicals that may prove toxic to the environment. Some of them contain trace amounts of heavy metals or hydrocarbons. Food contamination through pharmaceutical products is mostly linked to animals. The pharmaceutical products are often used in treating various animal diseases such as infections. When drugs such as antimicrobials are used in animals, they form part of animal waste and when released, it finds their way to food and water. The challenge with pharmaceutical drugs is that there has been a growing resistance to the use of such products. For instance, with antimicrobials, there has been a growing resistance that occurs through the continuous use of such products. Some of the chemical components in pharmaceutical drugs are very harsh and more farmers are looking towards to use of natural herbal products in treating various livestock diseases. Dumping of pharmaceutical products such as gloves and other products has also contributed to the increase in chemical contamination.

Radioactive elements

Radioactive elements are chemical elements in nature most of which do not occur naturally. They are brought about by exposure to certain soils during mining or through soil contamination. Some of the radioactive elements such as uranium are used in large-scale production of electricity. The challenge with radioactive elements, their wastes are usually not easily disposed of. It takes a long while before their radiations die out. Food exposure to radioactive elements is through explosions of chemical reactors and use of nuclear weapons. For instance, in Japan, there were problems with the Fukushima nuclear power plant that led to its damage and saw food stuff being contaminated. This called for frequent monitoring of water and food samples to ensure they are safe for consumption. It also led to the setting of certain laws that would ensure minimal risks of exposure to radioactive elements. Radioactive element contamination can also occur in seas through the release of these elements in water bodies in the form of waste water. Some of the radioactive wastes slip from the storage facilities and drain themselves into soils and ground water leading to heavy contamination. Due to the nature of the risk of radioactive contamination, several countries have put in place various risk assessment strategies to ensure there is no radioactive contamination more so in drinking water supplies.

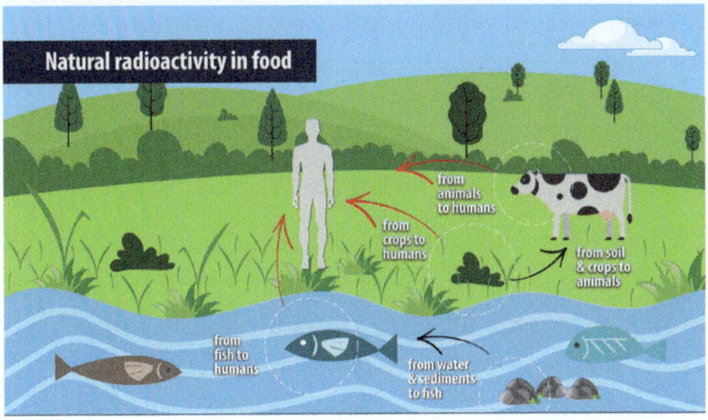

Figure 4.3. Radioactive contamination of food has the most severe effects on human life.

Source: (Infograph: F.Nassif/IAEA): https://www.iaea.org/newscenter/news/natural-radioactivity-in-food-experts-discuss-harmonizing-international-standards

Chemical contamination through electronic waste

Various devices have beem used and disposed in the current technological world. Such disposed electronic wastes, however, are not properly managed and have over the years become an issue that needs to be addressed. Most of the devices are made of chemical products that can lead to chemical pollution of soils and water. Some of the substances found in electronic devices include heavy metals and their metalloids, PCBs and dioxins among others. As these chemical components break down from the electronic device, they find their way into food and water (de la Guardia, M. 2021).

Chemical contamination through plastics

The use of plastics was quite common in the early 20th century as they were used in the packing of various products. However, due to the massive environmental pollution, it caused the use of plastics was banned in certain countries. Most of the already generated plastics have been left as wastes in rivers or dumping sites. Plastics are made using various chemicals that have been released into various beverages and foods. As plastics are often used in food packaging, it allows easy access for such chemicals to slip into such food products. The chemicals that slip into food are commonly known as migrants and have been found in bottled drinking water. The release of chemicals from plastics into food substances has been linked to various factors such as temperature and extensive storage of these elements. Exposure to such chemicals is often at times a low risk.

Contamination due to nanoparticles

Recent studies have shown that food contamination through chemicals can also occur due to the presence of nanoparticles. Nanoparticles are particles whose dimensions are very minute. These particles have over the years gained popularity and are used in making chemical products such as pesticides, cosmetic products and paints among others. More research is being done to properly evaluate the mechanisms by which nanoparticles get into food and the main pathway for contamination in human beings and animals. In most cases, wheat-based products have been found to contain large amounts of nanoparticles.

Detection

Detection of chemical contamination in food is achieved through the use of various tests done using various technological devices. Most of these devices are analytic in nature and are very sensitive such that they are able to detect the slight presence of chemicals in food stuffs. The devices are usually very sensitive.

Monitoring

Monitoring the presence of chemicals in food is very important as it is very easy for contamination to occur. There are various contaminants that could lead to chemical contamination of food. The extent of contamination determined whether the effects will be acute or chronic. Monitoring of chemical contamination can also be indicated by the type of symptoms presented by the individual. For instance, chemical contamination by lead is noted by the occurrence of certain cancers and retarded brain development in children. Monitoring can be done at two main levels that include the individual level and the community level. Individual level monitoring involves sampling of blood. The blood is checked for the presence of lead. The amount of lead found in blood is an indication of the extent of contamination by lead. Though blood sampling allows the detection of lead in the body, it may not be an indication of the presence of lead in certain parts of the body. The presence of lead in the body has been achieved through X-ray fluorescence. It allows the detection of lead in the entire body. Monitoring the presence of chemicals has allowed the determination of chelating agents. These agents have the ability to reduce the amounts of lead in the body.

The other type of monitoring is at the community level. This method of monitoring is useful in identifying the contaminated sites. This kind of monitoring is also useful in identifying the various health risks of various chemicals in soils. Monitoring is very essential in determining food safety. It also ensures that contamination is properly controlled.

Chapter

5 HOW INDUSTRIAL AGRICULTURE AFFECTS OUR WATER

Contents

5.1 Introduction 88

5.2 Agriculture as a Source Of Water Pollution 88

5.3 Negative Effects of Water Pollution From Agriculture 90

5.4 Rules and Regulations Placed Concerning Water Pollution and Agriculture 93

5.5 Environmental Justice, CAFOS, and Water Pollution 93

5.6 The Scope of the Water Quality Problem 97

5.7 Introducing Agricultural Fertilizers Into Water Supply Systems 101

5.8 The Future 103

5.1 INTRODUCTION

Agriculture is a sector that is practiced in every country because of the advantages it carries. Industrial agriculture is a new and more modern form of agriculture that involves the use of machinery that is more advanced for the production of crops and animals. It leads to increased production of crops and animal products hence helping to eliminate world hunger, and mass production leads to decreasing food commodity prices.

According to research, in countries like the United States of America, industrial agriculture is one of the significant factors promoting water pollution. Researchers from the National Water Quality Inventory of the Environmental Protection Agency (EPA) a few years back discovered that many rivers and lakes in the United States are very bad. They found that most of them have a very high level of nutrients and algae accumulation hence making the poor water quality and not safe (King, B. J. 2021).

The effects of water pollution are felt not only by the environment but also by human beings and animals. For instance, when people consume such water, it can lead to diseases if not treated. The effects are worse felt by animals, more so aquatic animals, who once encountered much water, and most of them die. Water with high levels of nutrients, mostly from fertilizers, are a risk to such animals. Polluted water also affects human beings in various ways, which include: it may affect their source of food since when water is contaminated, fishing becomes tough. This is caused by the algae blooms and the death of aquatic animals. Recreational centers may close down, for instance, beaches, because of algae accumulation.

5.2 AGRICULTURE AS A SOURCE OF WATER POLLUTION

5.2.1 Industrial animal agriculture

Here, animals are brought together and reared on a large scale. This process involves using machinery and highly developed systems that enable the operation of animal keeping to be more accessible and get a lot of produce. When animals are concentrated in a particular place on a large scale leads to a lot of animal waste being produced daily.

Research has found that countries like North Carolina have made a lot of animal waste over the past years due to industrial animal agriculture. This waste comes from both cattle and poultry. The most effective way that has been used to get rid of this waste is storing them in pits or open pounds called lagoons. When the weather changes, it rains, or a hurricane occurs, this animal waste can leak or rupture if there is a severe storm; other times, this waste is spread on farms like manure for crops. In case of a leak, burst, or spraying on the farm, such animal waste may have adverse and severe negative effects on the environment. These effects may be very devastating (Kimbrell, A. (Ed.) 2002).

The effect of these may include contamination of water that may lead to the death of aquatic animals hence reduction of sources of food, or adverse impact on human health when water is contaminated and injected by a human being because of having a lot of nitrates. The most harm comes from the fact that animal waste may include harmful substances like heavy metals, or animals under medication may excrete these in their destruction. The waste may also contain harmful bacteria and viruses, especially from animals like pigs. Aside from that, they may also include a high concentration of nitrogen and phosphorus. When this waste is spread on land to promote the growth of crops, not all is absorbed, and the excess may be washed away when it rains into streams of water, rivers, lakes, and oceans. A good example is chicken waste has a high concentration of ammonia, which may be toxic to fish and other aquatic animals when it is chemically converted to toxic nitrates, and they consume it. It may lead to their death in massive numbers (King, B. J. 2021).

5.2.2 Industrial crop production

In crop production, the introduction of harmful fertilizers mainly brings about water pollution. Increasing industrial crop production involves using modern machinery, manufactured fertilizers, and improved laboratory crops. The methods put in place led to massive production and new and resistant crops, reducing world hunger and starvation. As much as it has its advantages, it also has its disadvantages.

When industrial crop production takes place, fertilizers used include fertilizers with nitrogen and phosphorus that help promote growth. When farmers use these fertilizers, they most often use more than is required, and the excess is washed away once it rains into streams of water, oceans, and lakes. Having so many nutrients in water may cause and boost algae's fast growth, which causes the inhabitation of fish and other aquatic animals. When they consume, they may die because of the toxins in the fertilizers. When human beings drink this water, they may end up developing sicknesses. Recreational facilities like beaches may also be closed due to the accumulation of algae. Also, these may affect the activities of fishermen due to the lack of fish and fishing grounds.

5.3 NEGATIVE EFFECTS OF WATER POLLUTION FROM AGRICULTURE

5.3.1 Acidification, algal blooms, and dead zones

When there are excessive nutrients accumulate in aquatic sources, it may lead to the growth of plants in these water sources. These plants include algal blooms and water hyacinths. When these plants develop, they cause a lot of harmful effects.

Eutrophication, or the expansion of aquatic plants, is caused by high quantities of nutrients in the water from animal waste and contemporary agricultural fertilizers—eutrophication results in hypoxia, or oxygen-depleted water. Algal blooms in aquatic settings are dangerous if not regulated. Some HAB strains produce bio toxins that can cause illness in humans and kill fish and other aquatic life, while others deplete the oxygen in the water, resulting in "dead zones" where aquatic life cannot live.

Fig 5.2 Farm and water pollution.

Source: https://www.istockphoto.com/search/2/image?phrase=farm+pollution

The Gulf of Mexico eventually receives nitrogen fertilizer from Midwestern farm fields and animal waste runoff, which is one of the primary causes of the Gulf's so-called "Dead Zone," an 8,000-square-mile region with low oxygen levels where no fish can live. River phosphorus concentrations are among the highest in the country in areas such as Maryland's Eastern Shore, which is home to hundreds of chicken broiler operations. This relates to the estimated 228 thousand tons of excess chicken manure spread around the state. The Chesapeake Bay, drainage from the Delmarva Peninsula's vast chicken farms, is frequently plagued by toxic algal blooms and dead zones.

Ammonia from agricultural runoff can destabilize ecosystems by acidifying rivers and streams, affecting river and stream ecology.

5.3.2 Contamination due to the presence of heavy metals

In Industrial agriculture, there is an extensive use of metals like copper and zinc, which are added to the food of animals like pigs to help boost their growth and to help prevent diseases. Once these animals consume these metals, some find their way to their waste and the environment. In the presence of rain or storms, this waste is washed away to lakes, rivers, and streams, hence contaminating the water (Kimbrell, A. (Ed.) 2002).

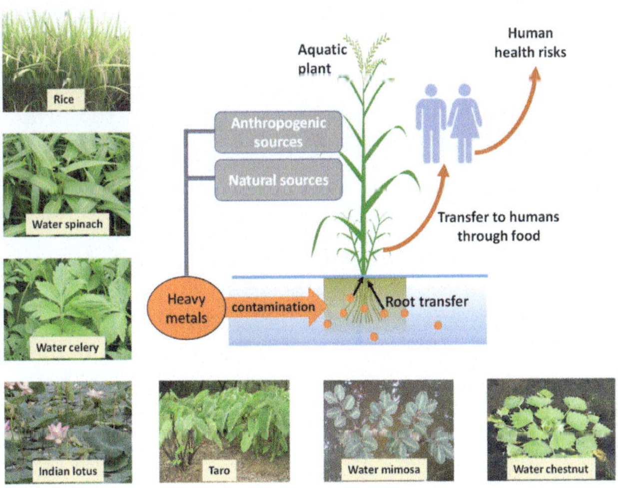

Figure 5.1 Heavy metals in agricultural catchments.

Source: Uddin MM, Zakeel MCM, Zavahir JS, Marikar FMMT, Jahan I. Heavy Metal Accumulation in Rice and Aquatic Plants Used as Human Food: A General Review. Toxics. 2021; 9(12):360. https://doi.org/10.3390/toxics9120360

When human beings consume this water, it may lead to diseases in their body as metals are not digestible by the human body. When consumed by aquatic animals, these metals may lead to their death and hence reduce the population of marine animals. For example, a metal such as zinc is very harmful to fish. Lastly, these metals help boost the growth of sea plants and water plants, such as water hyacinth, which causes the habitat of fish to be limited.

5.3.3 Nitrates in drinking water

High nitrate levels in drinking water can cause harm to people, for example, low birth weight and low oxygen levels. High nitrate levels in drinking water may indicate the presence of pesticides and other agricultural toxins.

Despite the researcher's findings that levels above three milligrams per liter of water indicate "pollution" and levels above one milligram per liter indicate industrial agriculture-caused contamination, the EPA imposes nitrate contamination criteria of 10 milligrams per liter to avoid blue baby syndrome. According to a state nitrate water pollution assessment, 53% of Delaware's water, 28% of Maryland's groundwater, and 10% of California's water had more than five mg/L of nitrates. All of these have a high concentration of industrial contaminants.

Another concern is the agricultural chemical pollution of drinking water. According to the California State Water Board, high amounts of uncontrolled cancer-causing chemicals are present in the drinking water of up to one million Californians' homesteads, most of whom live in Central Valley agricultural regions (King, B. J. 2021).

5.3.4 Outbreak of diseases and contamination due to pathogens

Waste from animals contains a very high level of disease-causing organisms, also known as pathogens. It is found that some destruction of certain animals has more than 100 pathogens and, once consumed, can cause illnesses in human beings. When it rains, or water washes away, these wastes go into water sources, rivers, lakes, streams of water, or oceans, it causes contamination of these water points. When people ingest this water, they may easily fall sick. Even during recreational activities like swimming, when human beings get to drink this water, it may cause waterborne diseases.

5.4 RULES AND REGULATIONS PLACED CONCERNING WATER POLLUTION AND AGRICULTURE

The Environmental Protection Agency (EPA) issued the first regulation addressing CAFO waste, as required by the Clean Water Act, in 2003. This restriction applies to all industrial farms since they are deemed point sources of pollution. If they posed a risk of damaging water sources, all CAFOs were required to get a National Pollutant Discharge Elimination System (NPDES) permit. The NPDES permit specifies discharge limitations, maximum bacterial counts, and other monitoring and reporting requirements.

Due to legal challenges, notably by the National Pork Producers Council, many changes were made to the original legislation. Under the new legislation, only large companies that dump manure directly into waterways must get a federal NPDES permit. Permits are unnecessary for facilities that aim to handle trash in lagoons before spraying or scattering it across fields as a disposal method. This is true despite these locations' regular leaks, spills, and runoff. Untreated runoff from rain or over-application is classified as agricultural stormwater, which does not require permits because it is handled as a nonpoint source. In recent years, many citizens have turned to litigation to protect their drinking water and local waterways from contamination by CAFOs. This is in response to the government's inaction and insufficient regulations surrounding CAFOs and water contamination. Despite public outrage, municipal and state legislators have passed laws that make it easier for CAFOs and processors to operate in surrounding regions, sometimes in collaboration with "Big Meat" firms. Furthermore, like many other states, North Carolina has a disproportionately small white population and a large concentration of CAFOs. The environmental justice movement has supported the struggle against CAFO contamination in certain areas.

5.5 ENVIRONMENTAL JUSTICE, CAFOS, AND WATER POLLUTION

In recent years, many people have turned to regulations to protect their drinking water and local waterways from contamination by CAFOs. This is in response to the government's inaction and insufficient rules surrounding CAFOs and water contamination. Despite public complaints, municipal and state legislators have passed laws that make it easier for CAFOs and processors to operate in surrounding regions, sometimes in collaboration with big firms. Furthermore, like many other states, North Carolina has a tiny white population and a large concentration of CAFOs. The environmental justice movement has supported the struggle against CAFO contamination in certain areas.

5.5.1 Pollution and regulation due on well water

Even though most Americans acquire their water from water sources, many individuals still rely on water from deep underground wells. According to researchers, 13 million households rely on private well water as their principal drinkable water supply. The Environmental Protection Agency (EPA) mandates municipal water systems to fulfill specific drinking water quality criteria, but it does not regulate or compel private wells to achieve these same standards. Private Wells and public water systems are generally not examined regularly. Households who use private well water must do their water testing to ensure the safety of their drinking water. As a result, they are susceptible to the negative impacts of industrial agriculture and other types of factory farming, which pollute the water.

5.5.2 How water is used in agriculture

In countries and states like the United States, agriculture consumes 80% of the country's freshwater. Most farms in the Midwestern United States use extended, overhead sprinklers that spin around a central axis to irrigate their crops. Center-pivot irrigation and related techniques stimulate the use of vast volumes of water. The Ogallala Aquifer stretches from Wyoming and South Dakota to the Texas Panhandle and gives water to about one-fifth of the country's wheat, maize, and beef cattle, up to 60% drained in certain areas. According to the USDA, climate change and existing water consumption habits will significantly reduce the amount of water available for agriculture in all US locations by 2060. According to research, seven of the ten most significant grain crops will reduce long-term production.

The United States relies on over 330 million acres of farmland to produce surplus food and other agricultural-related things. The agricultural sector in the United States is well-known internationally for its high output, superior quality, and excellent customer service. However, water quality might be affected if farms and ranches are not appropriately maintained. According to the 2000 National Water Quality Inventory, agriculture is the primary cause of nonpoint source (NPS) pollution in water, the second leading cause of water quality pollution in surveyed rivers and lakes, and the leading cause of NPS pollution in research is livestock keeping.

Examples of agricultural operations that contaminate NPS include poorly located or mismanaged animal feeding facilities, excessive grazing, improper timing of plowing, and inaccurate, excessive, or erroneous applications of pesticides, fertilizers, and irrigation water. Silt, fertilizers, viruses, pesticides, metals, salts, and other pollutants are released due to agriculture and livestock keeping—agricultural practices impact surface water and groundwater, which may be mitigated using locally-tailored management measures. (Kimbrell, A. (Ed.) 2002)

Nonpoint source pollution: Unlike pollution from concentrated sources like industries and sewage treatment facilities, the pollutants that contribute to nonpoint source (NPS) pollution come from several dispersed sources. Rain and snowmelt that seep into the earth can cause polluted runoff. Runoff collects and transports natural and anthropogenic pollutants, which are then discharged into water points via lakes, rivers, wetlands, oceans, and aquifers.

They are saving money on environmental costs while increasing output is a win-win for farmers and ranchers. Several government programs are designed to help farmers and ranchers develop financially feasible management options for preventing and addressing NPS contamination. For instance, over 40% of the Clean Water Act's Section 319 funding has been utilized to mitigate natural resource site (NRS) pollution from operational farms and ranches. Additional financial incentives, technical assistance, and cost-sharing are available from various state and nationally-backed programs to assist with implementing pollution management measures in the NPS. Many groups and people in the area have worked together to create regional support networks to use technologies and strategies to eliminate the impacts of agricultural processes on water quality.

Sedimentation: Transported dirt from farms is the primary source of water pollution in agricultural settings. Silt is carried by rainwater and ultimately discharged into surrounding lakes and streams. Reduced light for aquatic plants directly results from silt accumulation in water. It also can obstruct the gills of fish and kill off fish larvae. Soil particles laden with pesticides, fertilizers, and heavy metals are often carried into water sources, which exhaust oxygen levels and trigger algal blooms that kill off most aquatic life. By controlling the quantity and velocity of runoff water, retaining soil, and stopping soil transfer, farmers and ranchers may cut down on erosion and sedimentation by 20-90% (King, B. J. 2021).

Nutrients: Nutrients like phosphate, nitrogen, and potassium are added by farmers using chemical fertilizers, manure, and sludge. They could plant legumes and dump fertilizers to boost yields. Nutrient sources are used immediately before rainfall or when plant demands are exceeded so that nutrients may sip to aquatic habitats. The presence of algae in the water can cause problems for recreational activities like boating and swimming, alter the flavor of drinking water, and even kill fish by reducing the amount of oxygen in the water. It is known that excessive amounts of nitrate in drinking water can cause illnesses like "blue baby syndrome" in babies. If farmers want to stop their crops, they need to start using nutrient management techniques that will help them keep their harvests at a high level while cutting down on fertilizer (Kimbrell, A. (Ed.) 2002).

Where animals are fed: Cattle sheds are small enclosures used by farmers and ranchers to feed and care for their livestock better. However, even small areas may become significant contributors to animal waste. About 238,000 farms and ranches in the United States generate 500 million tons of manure annually as part of animal feeding activities. Nutrients, oxygen-demanding organics and sediments, bacteria, viruses, and other diseases may be carried from poorly maintained facilities to water sources, causing harm to fishing zones. The quality of groundwater may be ruined if sewage leaks into the soil. Farmers and ranchers may reduce environmental pollution by using efficient waste management systems to control wastewater and pollutants from their farms.

Domesticated Grazers: Overgrazing reveals soils, speeds up erosion, destroys fish habitat due to sedimentation and silt disposal on water sources, and threatens the life of the plants and animals that live in and around streams and water sources, which are extremely important to the maintenance of healthy ecosystems and the excellent water quality. Farmers and ranchers may reduce the negative impacts of grazing on water quality by varying grazing intensity, keeping animals out of sensitive areas, providing alternate water and shade sources, and encouraging the re-vegetation of ranges, pastures, and riparian zones.

Irrigation: This is supplementing rainfall with irrigation water or using irrigation water to keep plants from drying out entirely. Poor irrigation systems might affect the quality of the water. Evaporation of irrigation water can lead to high salt concentrations in arid places when rainfall cannot reach the soil depths and bring nutrients. Erosion, fertilizers, pesticides, heavy metals, and a decrease in the quantity of water that would otherwise flow through streams and rivers are ways excessive irrigation can severely affect water quality. In addition, selenium can build up; this is a poisonous substance that can harm duck reproduction. Reduced irrigation water usage can help farmers reduce NPS contamination. They can accurately predict crop water needs and water only as much as is required. Some farmers may upgrade their outdated irrigation systems with cutting-edge automated systems.

Pesticides: Many chemical pesticides, including fungicides, insecticides, and herbicides, are employed in agriculture. These contaminants can enter water supplies through various routes, including direct contact, drainage, and atmospheric deposition. They threaten food security, aquatic life, and wildlife habitats. Farmers should use Integrated Insect Management (IPM) techniques tailored to the specific soils, climate, pest history, and crop circumstances of each field to reduce the likelihood of chemical drift. Sustainable pest management is achieved by the establishment of natural barriers, the management of pesticide usage, and the regulation of critical applications.

Farm Bill Conservation Programs: The Farm Bill, signed into law by President Bush in May 2002, allocates up to $13 billion in funding for conservation initiatives over six years. The funding available for conservation projects to reduce polluted runoff has been increased by 80 percent thanks to the Farm Bill (King, B. J. 2021).

5.6 THE SCOPE OF THE WATER QUALITY PROBLEM

Agricultural pollution sources have been increasingly prominent since the mid-twentieth century. This is due to increased agricultural output and decreased wastewater released from human settlements and industrial sources. As a result of farming activities, there is an increase in the release of nutrients and other pollutants (such as pesticides) into the aquatic environment.

This is due to poor soil management and fertilizer application, biological pollution (such as that produced by microbes in manure), and agricultural soil erosion. It is estimated that farming contributes 60% of nitrates and 25% of phosphorus to water bodies in the United Kingdom and 75% of sediments that lead to water pollution. Sixty-five percent of Scotland's water bodies are seen better, but agriculture is blamed for the degradation of the remaining 35 percent. Finally, just 22% of Northern Ireland's water bodies are classified as okay. Water management issues are a vital concern in the agriculture industry. Agricultural pollution may be caused by several or a single point source, such as a poor storage facility.

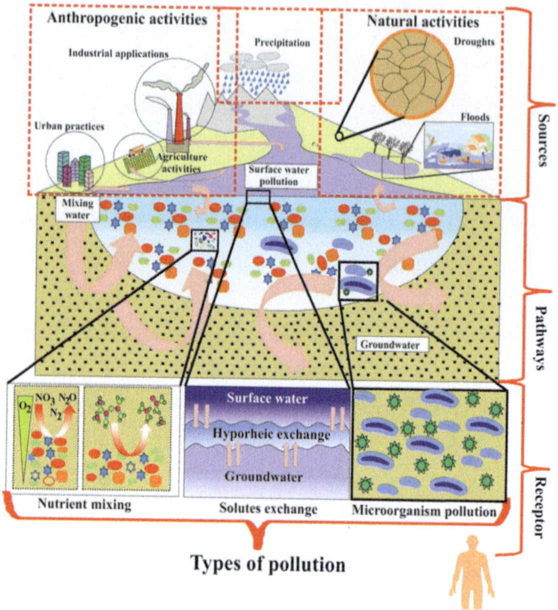

Figure 5.2 Farm pollution explained.

Source: Akhtar N, Syakir Ishak MI, Bhawani SA, Umar K. Various Natural and Anthropogenic Factors Responsible for Water Quality Degradation: A Review. Water. 2021; 13(19):2660. https://doi.org/10.3390/w13192660

As a result of pollution's adverse effects, water nutrient concentrations may rise. The complexity of the processes by which nutrients and pollutants leave the land, including an interaction between locality (such as slope, rainfall, and soil type), management (such as plowing, input regimes, field margin management, and so on), and a time lag, can make identifying pollution sources difficult. A more in-depth understanding of field-management water interactions is necessary to teach farmers how to adapt agricultural operations to reduce these impacts while maintaining farm profitability.

Climate change may change the water volume, water flow patterns, and the movement of toxins from farming areas or regions to water sources. To control these disadvantages, future laws must be set that must require proper land management. Climate change may have an impact on future agriculture production. Establishing agricultural operations with minimal effects on water quality is challenging. Water policy in the United Kingdom is implemented on several dimensions, from regional to national to European, and even down to sub-catchment, water body, farm, and site levels. When businesses in the United Kingdom can produce an understandable message, it may be simpler to engage in collaborative actions to enhance communication across several levels of water policy.

When combined, the Water Framework Directive (WFD) and CAP subsidies have enormous potential to encourage sustainable agriculture practices across Europe. According to the data, poor implementation and targeting contribute to the ineffectiveness of many regulatory or stewardship initiatives. Various regulatory measures have been adopted or imposed by the devolved governments and England to mitigate rural pollution. Nonetheless, there is a need to improve sector and scale alignment for these activities (Kimbrell, A. (Ed.) 2002).

Excellent examples of good farm management practices exist (such as not using manure and fertilizer until after heavy or prolonged rainfall, incorporating manure into the soil as soon as possible, and using slurry injection techniques). Still, they are only used in a few regions. This shows that there is still room for improvement in water quality. For example, if fertilizer prices rise, farmers may reduce their use of fertilizers due to the financial hardship it would create. To increase farm management efficiency, it will be required to identify treatments that drive behavioral changes. Improved weather forecasting would assist the environment and the economy since farmers could apply fertilizer appropriately. Adapting to changing conditions necessitates creative thinking. As a result, increasing the acceptance of specific techniques and encouraging the adoption of innovative agricultural applications are additional challenges. In this regard, demonstration projects have been successful and may benefit from further collaboration in the future. The United Kingdom now lacks a framework for transforming data into policy and actual change in agriculture, the environment, and water. As a result, further effort is required to address this risk. A free advisory center for information sharing might be developed to give a consistent message backed by research, policy, and demonstration projects to stimulate the spread of best practices. Because the United Kingdom is geographically diverse, its various areas have varying effects on production and the environment. "Smart" landscape design, which takes full advantage of the surrounding environment, may also help provide ecological services such as water and biodiversity and increase agricultural output. Water quality monitoring throughout time may help to understand environmental processes and their consequences better. New data retained for open user access, as well as advancements in modeling abilities, will allow for a better understanding and mapping of alternatives and threats, as well as the interaction between food management and its effects on water (King, B. J. 2021).

Dissolved and suspended inorganic and organic substances make up one-fifth of water. Because natural waterways' chemical and physical composition varies widely, freshwater ecosystems have adapted on a regional scale to address the specific difficulties they provide. Because they originated in response to specific environmental circumstances, ecological networks can be susceptible to introducing chemicals into the environment. They can undergo fast changes when substance levels shift. Water quality may be assessed partly by comparing the current state of the water to its normal/natural state because pollution is typically defined as a significant deviation from normal or "natural" chemical conditions, usually due to human activity. Any myriad of dissolved or particulate compounds that exist naturally or are synthesized can be used to assess water quality.

In the first two-thirds of the twentieth century, the poor quality of fresh water in UK watercourses was mainly caused by effluent from industrial sources and human habitations. However, the relative importance of the various pollution sources has recently shifted. Environmental regulations for point-source emissions have become more stringent. In contrast, technological advances in production have lowered the number of contaminants in industrial effluent in the United Kingdom (water pollution coming from a single point). Larger, more productive farms have resulted from a significant increase in the world's agricultural land. Agriculture affects water quality because of nutrients and other chemicals released into the environment from soil management and fertilizer application, biological pollution caused by microorganisms in manure, and soil erosion and farmland runoff caused by farming activities.

Environmental changes can influence river water quality. Agriculture may affect water quality in two ways: directly and indirectly. The soil erosion, fertilizers, and pesticides from farms into streams during rainstorms immediately impact the ecology. Consider upland drainage, intended to promote grassland, as an example of a measure with a secondary impact. Rapid river flows, followed by bank erosion, may worsen sediment issues downstream if drainage accelerates the rate at which water is lost from the slope following rainfall. Reducing plant cover on agricultural land adjacent to river margins and banks makes it possible to enhance river water temperatures and the capacity of river water to hold dissolved oxygen. These activities have significant direct and indirect effects on in-stream ecosystems, such as increased nutrient enrichment and its negative repercussions (eutrophication).

5.7 INTRODUCING AGRICULTURAL FERTILIZERS INTO WATER SUPPLY SYSTEMS

The presence of fertilizers is a significant contributor to water pollution. Eutrophication is caused by all types of nitrogen and phosphorus, which encourages the growth of algae and other undesirable aquatic organisms, such as poisonous algae. Agriculture is responsible for 60% of the nitrates and 25% of the phosphorus found in UK water bodies.

The erosion and movement of soil particles and fine silt from agricultural land into rivers can impact fish spawning and the quantity of light available in the water. Because it disrupts natural processes, this may reduce the river's flow capacity and raise the danger of flooding. Because some pollutants (especially pesticides and phosphate compounds) readily attach to the soil, they may be transferred to surface water via this route. Farming is responsible for 75% of the sediments that pollute water bodies. As a result, the agricultural and rural land management and wastewater treatment sectors are the leading contributors to sediment-related water quality concerns in the United Kingdom's WFD River Basin Management Districts. Organic chemicals and fertilizers are two of the most frequent categories of agricultural chemical toxins. Their breakdown products are often exceedingly dangerous to aquatic organisms and have several adverse side effects. The European Union, on the other hand, has prohibited the use and distribution of some persistent and bio-accumulative pesticides. The European Food Safety Authority (EFSA) conducts a rigorous risk study on all pesticides to filter out chemicals having these properties. Pharmaceuticals from veterinary drugs entering streams and their impact on ecological processes are also significant issues. Although agricultural sources are significantly less prevalent than sewage effluent sources, recent advances in this sort of study show that pharmaceuticals are ubiquitous in our waterways. There haven't been many efforts put into determining the cause of the problem or the full scope of its implications. Agricultural microbiological pollutants are the most prevalent source of water pollution from diseases such as *E.*

coli, Cryptosporidium, and Campylobacter. Since the 1960s, the European Union's Common Agricultural Policy (CAP) has been a critical economic and agricultural force. Currently, the Common Agricultural Policy (CAP), a farming assistance program that administers 50% of the EU's land area, receives more than 40% of the EU budget. The great majority of farmers now rely on government assistance. Massive increases in nitrogen input into the environment followed the 1970s oil boom when relative nitrogen fertilizer costs fell dramatically (King, B. J. 2021). However, this tendency began to decline in the 1980s as environmental concerns began to limit nitrogen usage. To solve the so-called "nitrate issue," the European Union Nitrates Directive regulates the amount and frequency with which nitrates can be administered (e.g., manure and fertilizer). Catchments in the United Kingdom that have surpassed or are on the verge of exceeding nitrate standards are subject to the Act and classified as Nitrate Vulnerable Zones (NVZs). Farmers who are found to be in breach of the NVZ restrictions may forfeit their CAP payments. Despite substantial limits on nitrogen intake due to nitrogen's long transit durations in water and the inability of some groundwater sites to complete denitrification, nitrate concentrations in groundwater remain excessive, even in protected areas. Water bodies are considered in "excellent ecological health" if they are at least 80% free of pollution and contain a biota typical of their region if they cover more than 24% of the total area.

On the other hand, agriculture is a crucial cause in 35% of water bodies, even though 65% of water bodies in Scotland are classified as outstanding or above, compared to 22% in Northern Ireland. Population growth and development, the economy, and the environment are all elements that must be addressed in the present and future of agricultural output. Fertilizers, pesticides, microbiological contaminants, and silt have been found in high concentrations in rivers and groundwater, but no management technique is to blame. A range of factors, including how the land is maintained and complicated natural features such as slope, rainfall, and soil type, can influence how nutrients and pollutants leave the earth (plowing, input regimes, field margin management, and so on). Pollution spreads throughout the country, with various sources, mobilization techniques, and distribution networks in different locations. Due to breakthroughs in environmental science, many of the physical processes and issues involved with land management and water quality in a smaller catchment area are now becoming clear to us.

However, these achievements pave the way for creating solutions to fight environmental water quality concerns while permitting ecologically responsible farming practices. Measuring the success of therapy, on the other hand, remains challenging. As noted in the European Commission's report on the WFD and catchment management, there should be a clear plan describing the required and necessary procedures that all farmers are expected to undertake, in addition to the additional supplemental measures that can be supported. It also recommends collaborating with the farming industry to ensure its success and broader adoption. A clear starting point from which all farmers may be informed is required to offer proper advice and enforcement.

5.8 THE FUTURE

Water scarcity is led by changing climatic conditions (e.g., higher temperatures), precipitation (amount, intensity, and seasonality), river flow, groundwater recharge, plant physiology (e.g., responses to elevated atmospheric CO2 concentrations altering plant water-use efficiency), and other factors. Several or an increase in heat or drought stress will negatively influence all types of agricultural output. If soils dry out and animals demand more water, it may be essential to increase irrigation. Farm management, water flow volumes, water movement patterns, and the associated transfer of toxins from agricultural land to aquatic bodies can impact water quality. Climate change has a worldwide economic impact of 10 percent. As the worldwide trade in agricultural commodities expands dramatically, economics in places that can meet the rising demand for food will face intense pressure to do so. Global warming may impact the United Kingdom less than other countries. As a result, future production is expected to improve, necessitating innovative measures to support agriculture while lowering its impact on water quality and other environmental services. Understanding agricultural system leverage points, such as landscape aspects where activities may be concentrated for optimum profit, is critical for finding a solution to this challenge. The agriculture industry in the United Kingdom might be secured for the future by drastically reducing costs via strategic planning. Instead of focusing on predefined goals, we should look forward and recognize that the world will swiftly alter. Even with the increase in food demand, we should not anticipate that agricultural intensification will continue globally.

In some cases, it may be feasible to establish sustainable food and water systems using cutting-edge technology, permaculture, and intercropping. Agriculture may take several forms, ranging from intense farming to more rural environments. If we wish to have resources in the future, they should not be depleted faster than they can be replaced. Depletion of natural capital (such as soil quantity, quality, or biodiversity) that limits future generations' chances is likewise undesirable. Future supply stability is critical to economic sustainability, allowing businesses to expand and help communities while reducing government subsidies. We must determine how to control and reduce the effects of variable agricultural harvest sizes.

The transfer continuum approach, a basic conceptual framework, may be used to explain the impact of agricultural management on water quality.

Fertilizers, soil amendments, and animal feed are all possible material sources. All of these processes occur during the mobilization phase, which begins when the material leaves the field and begins its trip. Solubilization is intricately tied to the soil nutrient cycle since it encompasses both geochemical and biological soil processes such as desorption and enzyme hydrolysis. Physical processes such as surface separation produce this separation. The third image depicts the nutrition transfer loop. A heavy downpour has weakened the ground. Incidental losses include crop losses caused by recently applied fertilizer or manure washing into hydrological routes before it had a chance to acclimatize to the soil. Substances must be delivered in a method that allows them to go from their mobilization location to the surface waters to be effective. To transport water, it passes through surface or subsurface routes that vary in both space and time. When the soil is saturated, or the intensity of rainfall exceeds the pace of soil penetration, water-containing pollutants may flow over the ground. Even though the source-mobilization delivery continuum technique was developed for phosphorus, it may now be used to understand and mitigate the effects of a wide range of other distributed contaminants. Because of the complexity of the contamination, tracing the source of diffuse pollution can be challenging. According to the continuum method, the effects of agricultural point or diffuse pollution may occur at a considerable distance and with a lag period as long as the pollutant is mobilized and conveyed through the catchment to accumulate downstream. A series of minor difficulties upstream may compound into a large problem farther down the line. Low-input farming upstream is likely to have contributed to nutrient burdens downstream, but this is a hidden source that is difficult to pinpoint.

The most efficient way to deal with diffuse pollution is to take a preventative rather than a reactive strategy, establishing best practices on farms to avoid localized impacts on fields that can have far-reaching consequences for water quality. For food-land-water management to be effective, upstream and downstream must be connected holistically and systems-based. More and more organizations are collaborating to solve this issue at the catchment level. The source-mobilization-delivery-impact technique may investigate the scales and interactions between land management and water quality. Source management, which analyzes all inputs and strives to optimize nutrient consumption efficiency, results in less environmental nutrient loss. Applying fertilizer at the proper time for the crop is an example of good source control. The farm's inputs, including bagged fertilizer, concentrate feeds, the atmosphere, and weathered soil, must be included when calculating nutrient intake. Plowing techniques that increase infiltration capacity and decrease soil erosion, as well as manure management procedures that reduce the possibility of leakage, are covered by mobilization control, which seeks to prevent soil or nutrient loss in the field. As a result, keeping soil and manure increases agricultural economic worth, and excellent soil and manure management enhance water quality. Slurry injection techniques, incorporating manure into the soil as soon as possible after application, and avoiding applying manures and fertilizers before heavy or prolonged rainfall are all good agricultural practices that can reduce the risk of nutrients reaching water bodies. This arrangement may be termed a win-win since effective nitrogen utilization saves farmers money while improving water quality in the long run. There might be further benefits, such as better crop growing conditions, larger yields, and fewer trips across the field to maintain soil structure.

Chemical leaks, septic tank output, and poorly maintained farmyard infrastructure are all visible and manageable at the site or field level. Agricultural technology must be enhanced to detect and handle these risks. Water may be transferred from farms to lakes and rivers in various ways. Water moves on the surface, beneath the surface, and even deeper below the surface through flow (within the soil) and deeper groundwater flow. Much of the energy of the overland flow is required for particle and colloidal detachment (mobilization), albeit this may help in their progress along the flow channel. In temperate nations such as the United Kingdom, flow, or water that has percolated through the soil and drained into watercourses via small subterranean channels or longer pathways via deeper groundwater, is the principal source of stream flow. Overland flow is more likely to occur when it rains for an extended period, the ground is saturated, or there is heavy rainfall (infiltration-excess overland flow). Particularly vulnerable sites include the foot of hill slopes, tractor or animal trails where the soil surface has been compacted, and thin, poorly drained soils that become wet more quickly. If climate change estimates for increasing precipitation intensities come true, future overland flow in the United Kingdom may rise. Ponds may be used to collect sediment, and buffer strips can collect nitrates and improve soil water denitrification via biological activity, all of which help to delivery control. It is critical to emphasize the favorable impacts of land drainage, under-field drainage, and planned channel alterations on productivity and the length of time land may be used. There is disagreement about whether drainage aids or hinders attempts to reduce dispersed pollution. Agricultural pollution can occur from discrete sources (such as a slurry storage facility). Still, it is estimated that the vast majority of agricultural pollution comes from more diffuse sources, making identifying and attributing specific activities or places more challenging. Identifying the direction pollutants travel and whether or not this movement may be disrupted is critical for preventing aquatic ecosystem contamination (Kimbrell, A. (Ed.) 2002).

Chapter

6 AGRICULTURAL POLICY TO CONTAMINATION

Contents

6.1 Introduction 108

6.2 Establishment of Food Safety Standards and Regulations 109

6.3 Creating Incentives for Food Safety Innovation 111

6.4 Establishment of Liability Standards for Food Safety and Contamination 113

6.5 Environmental Impact of Agricultural Based Contamination and Policies Related 116

6.6 Dangers Facing Policies in the Agricultural Sector Relating to Contamination 117

6.7 Benefits of Certified Food Management Systems 119

6.8 Benefits of Food Safety Training For Employees 120

6.1 INTRODUCTION

In too many states, agriculture remains at the heart of economic growth and stability. It lays a fountain of food security and more-so, food safety. Such are the acts that define the arguments behind the genetically modified organisms commonly referred to as GMOs just to mention but a few. This is then termed as agricultural policies and may vary from country to country. It describes the framework of government-implemented interventions and incentives used to address the issues facing agriculture. The competitiveness, productivity, profitability, and well-being of farmers and their families are all being improved by these varied policy approaches.

However, the primary function of Agricultural policies is to ensure food security, a factor directly hindered by contamination relating to agricultural activities. Contamination directly interferes with the economic and social welfare of a country's people. Food safety is imperative to maintaining a healthy and stable food supply. The ability of farmers to produce safe, high-quality crops is threatened by contamination from soil, water, and air (De Leo, F., & De Donno, A. (2018).

The need for agricultural policies on contamination is fueled by human evolution. Over the years, agriculture has evolved from subsistence farming to what it is today with an emphasis on commercial farming, food processing and distribution. It is relatively termed as a political aspect. It arguably deals with the management of rural development, food production and nutrition in an environment that supports sustainable growth with the least possible damage to future generations. The agricultural policy required by this modernization and development attempts to provide people with the knowledge and skills they need to adapt to their environment. It also offers a way to end poverty, particularly in rural areas where it is most prevalent.

Agricultural policy enactment especially when it comes to contamination is done by governmental bodies such as the CDC. Maintaining a sustainable balance between food supply and its demand requires adequate attention by governments at local, national and international levels more so in regions where agriculture is the main source of individual and public revenue. To attain maximum economic benefits and environmental sustainability, welfare, and development of quality of life, it is important to use biological inheritance, natural resources, and the environment responsibly. To prevent any potential harm or contamination of our environment by agriculture, regulations and policies are developed in accordance with the principles of sustainable development; this will also facilitate social processes within communities and cultural exchanges with different ethnic groups. The urge to make big money in today's world has spurred a lot of industrial activities over the recent past. By this, chemical and industrial pollution has been on the rise. Take for instance the rising use of inorganic fertilizers. The growth in the use of fertilizers has indeed spurred agricultural production to suit the rising human population. This is however not the absolute case. While providing important nutrients and fiber to the diet, it also causes pollution through industrial and commercial activities unrelated to food production. The main concerns relate to soil erosion, pesticide contamination of surface water bodies, air quality problems, and water pollution due to fertilizers and livestock waste, greenhouse gas emissions from equipment.

But yet, these policies do not necessarily aim at banning or eliminating modern gains in the field of agriculture. Instead, agricultural policies on contamination are meant to control the use of such products such that they do not become hazardous to human life. As a matter of fact, it is imminent that farmers today apply such products to feed and to make money. However so, the cost of pesticides and fertilizers has become high, costs have risen, and the supply of these inputs is increasing. The poor quality of soils due to excess industrial runoff, pesticide drift and water scarcity has led to declining yields and decreased productivity—to put it simply, farmers are getting poorer (De Leo, F., & De Donno, A. (2018).

Such factual data help control the said agricultural policies and align them such that the farmers in question are not negatively impacted. It reflects the context of agricultural policy as a national food security crisis. The following are some of the Agricultural policy responses to contamination.

6.2 ESTABLISHMENT OF FOOD SAFETY STANDARDS AND REGULATIONS

The implementation of agricultural policy responses to contamination must promote food safety. Establishing rules and regulations for food safety is a crucial first step in boosting consumer confidence in the product. Food safety standards and regulations are literally policies in the sense that they are enacted rules passed into law aimed at protecting public health, reducing hazardous risks, managing contaminated materials in the agricultural sector and setting technical standards and guidelines.

Objectives of food safety standards

- The establishment of food safety standards and regulations helps consumers and farmers make informed decisions by ensuring they have a finite understanding of what is required during production and further to consumers on what has been used in the course of agricultural production. For example, the United States Department of Agriculture's (USDA) Food Safety and Inspection Service (FSIS) tracks the safety of approximately 20,000 products on a daily basis for several disease-causing microorganisms found in food. Through the FSIS testing program, consumers can know how foods were handled in production plants so they can be confident that their foods are safe if consumed.

These standards and regulations include proper labeling, food handling, cooking practices, and sanitation and inspection programs.

- Reduction of morbidity and mortality rates. Governments can prevent foodborne illnesses and diseases from affecting employees and consumers, who are more at risk from exposure to contaminated foods, by implementing certain guidelines and rules. It invokes that government bodies are entitled to coming up with the rules to ensure continuous monitoring of the quality of food products such that anyone consuming them can be confident of their safety. This will go a long way in reducing mortality as a result of food contamination.
- Boosting economic prosperity. Safety standards and regulations help boost economic prosperity by ensuring a safer supply chain. It further allows businesses to produce more high-quality goods at a lower cost, and by extension get more customers. By keeping a healthy society, then the community also becomes productive meaning greater business entrepreneurship and further external investments.

Promotion of food safety education and awareness. This is done through education, interventions, and control measures. The U.S. Food and Drug Administration (FDA) is responsible for reviewing the safety of all foods sold in the United States before they can be put on store shelves. They do this by setting guidelines, rules and regulations that businesses must meet when manufacturing or selling their products. This allows consumers to receive accurate information about what they are eating so they can feel confident in their purchases while still being able to know that they are not ingesting harmful chemicals or pathogens such as *E. coli* O157:H7-which could harm them or potentially make them sick (De Leo, F., & De Donno, A. (2018).

6.3 CREATING INCENTIVES FOR FOOD SAFETY INNOVATION

The nobility of financial psychology and infrastructure allows the governments to pay farmers such that they can take care of their plants and animals without involving themselves in chemical use and food contamination. One such example is the US farm bill enacted to pay farmers to take care of their animals. The farm bill enables the federal government to enter into more risk-sharing agreements with producers. It also establishes innovation zones in which local growers can submit proposals for new food safety systems. Incentives to innovation have been successful in other areas of the economy, such as energy production and alternative energy production, so why not agriculture? These incentives could be paired not just with funding colleges and universities to create food safety researchers but also with food safety regulations to help reduce problems like Listeria and other bacteria often found on produce.

Genetically modified organisms famously known as GMOs have also been applied by some governments in providing seeds to farmers for free before they can sell their produce back to the government. While this enaction has been controversial in the recent past, it has gained traction on matters of food production. In some countries such as Brazil, the use of genetically modified crops plays a significant role in reducing contamination by alien pests and helps create incentives for food safety innovation as an agricultural policy. Its widespread use and the increase in food safety regulations have created an incentive for innovation and technology to avoid potential contamination (Labrenz, M. (2022).

Concerns relating to the use of GMOs

GMOs are produced through the use of genetic engineering. Such foods are produced in laboratories using recombinant DNA techniques to introduce new traits into plants, animals and microorganisms. The mainstream media and public opinion have been concerned about the safety of GMO products, including various concerns over genetic mutations, transfer from one species to another (cross-contamination), presence of toxins and allergens, production of novel toxins and resistance to pesticides. Some of these concerns have been laid to rest by research studies showing that gene manipulations do not negatively affect animal or human health or gene expression in the body. Other concerns pertaining to off-target effects (detrimental outcomes unrelated to the intended target) may be valid, however current approaches for risk assessment rely primarily upon animal feeding trials that are similar in design to FDA mandated testing for other drugs.

Methodologies in creating incentives include;

- Providing tax incentives

Tax incentives for safe food production have also been a subject on the rise. They are necessary to promote safe food production and help farmers invest in biotechnology and other technologies to reduce their risks. On the same note, it is viable to engage hefty taxes on those producing foods with a risk of contamination. The size of the tax is always twice as great as what is paid on safe foods. Tax incentives are also a regulation by the World Bank. The World Bank has estimated that annual food losses due to contamination cause more than 1 million deaths annually worldwide. Preventing these losses requires incentive schemes that promote safer production methods and good practices, as well as regulation at all stages of the supply chain including storage, transportation and sale (De Leo, F., & De Donno, A. (2018).

- Providing tax breaks on safe food production.

A tax break is an incentive that reduces tax liability. This can be done by lowering the rate of tax on a product, but this only applies to high-profit agricultural products. In contrast, low-margin or risky farm products are not given any tax breaks and must pay their fair share of taxes as normal. In the United States, the federal government has two primary ways of supporting agricultural producers. The first is through direct payments, which helps farmers cover some of their operating costs. The second is through tax breaks. The first one helps keep farm income low, while the second creates incentives that encourage farmers to produce nutritious crops (e.g., fruits and vegetables) rather than commodity crops (i.e., corn and soybeans).

- Public-private partnerships

Public-Private Partnerships are a way for the public sector to partner with private firms. Through Public-Private Partnerships, the public sector can engage with and promote safe food production, improve access to safe and nutritious food, and better share the costs of supporting healthy diets with consumers. The public sector i.e. the government is responsible for funding infrastructure and information systems. The private sector responsible for food production will also be liable to obtain machinery in the developmental agenda. African countries have allowed the growth and development of irrigation schemes such that governmental bodies are able to promote irrigation and provide seeds not excluding fertilizers to farmers.

- Governmental grants and loans for companies that promote new and innovative ways of food production.

Governments have the ability of providing loans to farmers in need of aid in safe food production. Through grants and loans, there is hope to bring innovative ways of food production to life. It further promotes sustainable practices such as: growing natural local foods; increased efficiency; use of renewable resources; farmer's markets; agro-tourism; organic farming; high-quality products; etc. In the European Union, for instance, the incentive program has provided over 20 million euros to small and medium-sized businesses for activities in the field of agriculture and food production.

6.4 ESTABLISHMENT OF LIABILITY STANDARDS FOR FOOD SAFETY AND CONTAMINATION

In 2015, there was an argument in the United States concerning the use of meat and its production. It prompted the Food Safety and Inspection Services (FSIS) to enact liability standards for food safety and contamination (De Leo, F., & De Donno, A. (2018). The organization has a role to play in both establishing and implementing standards for food safety. The agency is responsible for setting standards for meat, poultry, eggs, processed fruits and vegetables, and non-food products like beverages, flour and spices. It also implements federal regulations on the storage of feed grains and seeds; oversees fish inspection; and ensures that animals in interstate or foreign trade are safe for human consumption.

In the recent past, other bodies have resolved to produce a certification program for all food handlers and food establishments. Certification programs help to

- Represent comprehensive knowledge, expertise, and competence to the public- Certification represents comprehensive knowledge, expertise and competence for the public to know about food safety laws, and how to comply with them. The certification demonstrates that you can protect the public's health.
- Assure consumers that the food handlers who prepare their meals

in an establishment have both adequate knowledge and/or training to prepare them safe foods.

- Establish best practices in the industry. Certification provides a structure to properly handle food and tableware. This ensures that all food handlers are being trained on food safety, sanitation and infection control. Better practices are also established through education and training, standards are monitored by the government, and penalties for violations become more severe over time.
- Providing opportunities for continuous improvement. Certification programs are designed to provide employees with opportunities for continuous improvement within the food industry. Over time, food industries with a certification program can become more proficient at their jobs and become role models for new developments in the agricultural industry. They may also develop an increased rapport with consumers and make more accurate food safety decisions, which is important to food establishments and their customers. While providing opportunities for continuous improvement, food industry employers can also help reduce liability (Labrenz, M. 2022)
- Maintaining common standards among food handlers. Certification programs help food industries gain access to terms such as foodborne disease prevention, safe food handling practices, and safe storage to enhance their capabilities in supplying high-quality food and beverages compliant with health code regulations.
- Developing public confidence in the safety of their food. Most food handlers in the United States for instance lack training, certification and records. Without compliance with regulations, it is impossible to ensure that restaurants and food establishments are operating safely. This also poses an economic burden to society as well as health risks to employees and customers. In order to protect public safety, there need to be mandatory training requirements for all food handlers, which can only be met with a certified program. Ensure protocols and procedures are followed. A worker who has had a food handler certification knows how to prevent contamination and can provide quality service to customers. One such example of the certification is the food safety

technician certification. The Food Safety Technician Certification will lead to a better quality of life by reducing food borne illness and preventing internal complaints due to the operational constraints which may have been restricted by liability concerns (De Leo, F., & De Donno, A. (2018).

Food handler certification programs provide the public with the assurance that the individuals obtaining certification are familiar with policies, procedures and practices related to food safety. In addition, employers can be assured that their employees understand the importance of maintaining a clean work environment as well as how to carry out their job responsibilities. At the same time, many violations of health and safety regulations are associated with uneducated workers or unskilled workers who have no knowledge of food safety measures. This can be avoided by ensuring proper training in all individuals associated with food handling operations.

Providing financial assistance for agricultural businesses to improve food safety practices

It is important that farmers are able to proactively manage their risks by implementing best practices throughout their operations. This policy will provide financial assistance so that businesses can implement these practices on a one time or regular basis to help ensure the safety of produce throughout their operation. It further increases food safety practices and reduces heat-related illnesses by providing financial assistance to agricultural businesses through outreach, education, training and technical assistance services to improve food safety. It will also enhance the capacity of companion organizations in rural communities that are located close enough to farmers but need additional resources to protect them from pesticide misuse.

Land tenure and land use

Land tenure and land use play a crucial role in the ability of a society to protect its environment through agricultural policy. Land tenure refers to any technique used by a landowner to govern their land. It can include leasing others their land, allowing others to occupy their land while they work on it and/or can involve selling off portions of the land but retaining ownership over it. Land use refers to the way that people allocate their resources, such as soil, water, good growing conditions and sunlight. Planning for a good agricultural policy includes shared decision-making between individuals, families and communities concerning the use of natural capital (acres) in environmental sustainability.

Economical impact

The economic impact of pollution can vary significantly depending on the kind of pollution, where it originates, and how it is expressed. For example, carbon dioxide emissions from countries that produce oil and gas may have a positive economic impact because they will be able to sell more goods and services. If an area experiences increased unemployment due to poor agricultural practices, however, its overall economy may actually suffer because fewer farmers have jobs that pay well enough to support a family. However, these policies have not taken into consideration their impact on the environment. Today, it is easy to see that land pollution, water pollution and air pollution are created by farm production (Labrenz, M. 2022)

6.5 ENVIRONMENTAL IMPACT OF AGRICULTURAL BASED CONTAMINATION AND POLICIES RELATED

Farming takes up a large amount of land, therefore it is important to protect the land from the effects of chemical contamination. The economic impact of farming practices is significant, and it is vital that agricultural policy addresses this issue within the context of environmental policy. The environment is not only being impacted by pesticides and fertilizers but also due to the conversion of land into agriculture. As farmers continue to increase production in order to achieve higher yields, these chemicals have found their way into our drinking water supply. This means that no matter where you live or what your source of water comes from, virtually all urban residents have this problem at some point (De Leo, F., & De Donno, A. (2018).

The toxic substances that are released into the air and water can be harmful to people and animals living close by, as well as people themselves. Studies have shown that agricultural runoff can cause many different health problems including asthma, cancer, gastrointestinal disorders, birth defects and poor mental development. These harmful effects are caused by toxic chemicals not just found in livestock waste but also pesticides used on crops. In order to maintain the environment, the agricultural sector has been encouraged by the government to take steps toward reducing soil erosion and chemical pollution.

The Agriculture sector generates about 14% of greenhouse gases and produces another 17% by other means. The UN has determined that agriculture can play a significant role in mitigating global warming, with the potential to contribute up to 5% of needed annual emission reductions. In order to help achieve this goal, agricultural organizations, including the FAO and IUCN, have identified the following priorities: increasing yields (to cut production impacts), improving land management practices (to reduce soils' carbon storage capacity) and developing methane digesters.

6.6 DANGERS FACING POLICIES IN THE AGRICULTURAL SECTOR RELATING TO CONTAMINATION

The agricultural industry is particularly vulnerable to contamination because of its close proximity to other industries and human settlements. Agriculture is also subject to heightened accident and emergency standards, as well as controls to protect human health. Through its policies on fertilizers, pesticides, animal feed additive substances, genetically modified organisms (GMOs), and other agricultural contaminants, the Government of Canada must ensure that the risks associated with these products are properly understood and managed.

Implementing food safety inspection and certification programs

It is the duty of all the members of the food supply chain to understand that ensuring food safety is a collective effort. Apart from the individuals within the supply chain that are directly involved with food handling, there are also those individuals that are indirectly linked to the chain. Such include the inspectors who provide certificates showing compliance to the food safety standards and the auditors. This aspect involving the food supply chain is basically there to ensure that consumers are protected from food poisoning and in some instances foodborne illnesses that can have a great impact on the quality of human life. Ideally, food safety certification is identified as a third-party verification process that all the systems in the food supply chain need to meet for them to be accepted as food safety standards. The certifications are normally based on tests, inspections and audits and tend to give the final consumer confidence in the products that they are purchasing. The third-party needs to be competent for the entire process to be successful. When companies have certifications, it normally inclines that they comply with the safety and fitness characteristics that are demanded by the various food standards. Most of these standards are brought about by the demand of the market.

Globally, food trade has increased and therefore increased concerns for food safety. For trade and better reputation of products and companies emphasis has been put on the issue regarding food safety. As a matter of fact, major food industry players that are involved with the management of food safety are insisting that regular inspections and auditing be done on the industries and insisting on certification. These certificates are only provided after thorough inspections have been done and proven to be conforming to the various standards (Labrenz, M. 2022)

Various food Certifications and what they mean for food industries

International Organization for Standardization (ISO) is a standardization organization that is responsible for the creation of international standards. As a food supplier, regardless of your scale or quality, it is your duty to ensure not only better management of the food product that you produce but also the health of your customer. This is the reason as to why ISO created various certifications to solve this problem. The food safety guidelines created by ISO help many organizations in identifying and controlling the food safety risks that may occur in an industry. Below are the various ISO certifications:

- ISO 22000- Food Safety Management System

This is a food safety management system that is applicable in companies all around the world from production to consumption. This is considered a major ISO Certification system that is important as it is beneficial for all individuals within the value chain. It is known to cover all the key factors that ensure food safety within an organization. It is considered a primary requirement in the business of food production.

- ISO 9001- Quality Management System

This ISO shows how much individuals responsible for the production or manufacture of food are dedicated to the production of high-quality food products and services. As a result of this, the clients get satisfied, employees are engaged in the production process, and there is organizational learning research, innovation and involvement of stakeholders in decision-making.

- ISO 14001- Environmental Management System

This ISO is applicable in all industries regardless of their size or functionality. This looks at ensuring that the safety of the environment is considered while production is taking place. The vision of the ISO is to create a framework that is recognized worldwide, and useful for risk assessment and performance monitoring to protect the environment. Take for instance industries with liquid effluents. Constructed wetlands are normally put up to help in cleansing the water that is getting out of the industries. Cleansing is done by natural treatment to ensure that there is no further damage to the environment. To be compliant to this ISO certification, it is the duty of the companies to do tests immediately after it has left the industry and after treatment. This is normally done to check the effectiveness of the process and whether something needs to be improved in the process to ensure that water is treated.

- ISO 45001- Occupational Health and Safety Management

This ISO looks at the Health of the employees in an industry and also the safety risks in the company. Apart from looking at the safety of food, the safety of workers is also paramount. Every individual that is handling food needs to be healthy and also safe to avoid contamination of the food product.

6.7 BENEFITS OF CERTIFIED FOOD MANAGEMENT SYSTEMS

The safety standards have become a big requirement by states around the world and also by the final consumers. Companies can benefit as stated below when they have a fully functioning and certified food safety management system (FSMS).

- Consistency- With a well-implemented FSMS, companies are able to produce safe and quality product thereby meeting the standards of the market thereby improving their profit margins.
- Due diligence- A well-implemented FSMS show that the company is meeting all its legal regulations and responsibilities when it comes to food safety regulation and legislation. Furthermore, it shows that it is doing so effectively. This will also help the company in case of court cases dealing with food contamination and poisoning.
- Consultation- with FSMS within the companies, all the relevant stakeholders normally work together with other employees within the company to ensure that all the food regulatory standards are adhered to. In doing so, they will be able to recognize and solve issues before they become larger problems.

- Liability- insurance companies tend to take their reputation to be the most important thing in their businesses. As such, there are those that will not agree to insure a company that does not have HACCP (Hazard Analysis & Critical Control Point) programs.
- Cost-effectiveness- when a company has a well-functioning FSMS, they will be able to notice cost-effectiveness given the fact that they will be able to diminish their contamination risks by costly food recalls.

Promoting food safety education and training

Training and education are important aspects in the life of an individual as it helps in determining between what one is required to do and what one should not do. The two basically act as stepping stones for safety as they link the theoretical aspects to the practical aspects. In theory, there are always good legislations that are ideally important in our everyday life. Taking for instance the food industry, implementing the good legislation means that all stakeholders involved need to undergo training and education. The shortcoming is that in most instances, they are normally not recognized at fair value. As such, due to poor design and implementation, the two normally do not meet the desired objectives.

6.8 BENEFITS OF FOOD SAFETY TRAINING FOR EMPLOYEES

Employees deal with food directly, their safety and knowledge regarding the importance of food safety should be treated with utmost keenness. As such, regular education and training regarding the same need to be conducted.

- Improves food safety

When conducting any workplace food safety training, the most paramount subject should be the prevention of illness. Pathogens are always considered to be of great concern. Chemical cleaning and pest control products may at times pose unseen danger and employees need to be informed of the same. In this training, employees are also normally taught about the various food allergens and the occurrence of cross-contamination. There are also different aspects of our bodies such as hair, dirt and fingernails that can get into the food causing contamination but with proper training, employees get how to prevent the same.

- Reduces food waste

It is not a new thing to hear the fact that companies tend to lose money every now and then due to food waste. This mainly occurs due to poor handling of the food, numerous spillages and also improper storage. With training, individuals will gain knowledge on the importance of proper food handling while reducing spillages.

- Boosts the reputation of your brand.

Consumers normally have high concerns regarding foodborne illnesses. In addition, tarnishing a company's image by giving its products a bad review is one thing that is not feared by consumers. These are some of the things that employees need to understand through awareness. With bad reviews, companies tend to lose money which then becomes a problem for employees who might lose their jobs. Providing them with proper training on food safety regarding how to avoid potential foodborne diseases, not only will the customers know that you care about them but it will also boost the brand reputation.

- Reduces workplace injuries and accidents

In most instances, chances are the employees in your company are also the consumers of the product you are producing. If it would get out that they are getting ill because of the same, the company will be in big trouble. This will be considered as a violation of the local safety violation.

- Compliance with safety regulations

It is a requirement for any food industry to always provide training and education to their employees regarding health and safety. As a matter of fact, companies get audited regarding the same and if found not to have accomplished the same, they would be required to provide a corrective action. In the case of the occurrence of an accident or an illness and noted that the company had not provided any sort of education or training to its employees, the business will be held legally responsible and will be required to cover medical and other costs. In the case where the accident or illness is severe, it may even force the company to shut down.

- Improves employee productivity

One way of motivating employees is through education and trainings. This is because through learning, employees get to be confident in what they do thus improving their productivity. On the other hand, with proper training, there will be reduced cases of accidents hence fewer absences, thereby ensuring that the workforce is strong and ready to face the challenges of the day.

- Reduces stress and saves money

With reduced lawsuits and increased productivity within the company, there will not only be reduced stress but also saving of money. Lawsuits are not only boring but also really expensive. As a matter of fact, in case where a company faces a lawsuit because of the dangers its product posed to its individuals, they tend to use a lot of funds and time to deal with the case. The shortcoming is that with the already tarnished reputation, the company takes long before it is able to recover financially. When business owners and employees are constantly worrying about potential food safety issues, it can be really stressful. Continuously doing trainings and educating employees is one way of preventing accidents, illnesses and lawsuits brought up regarding food safety. Furthermore, there is the reduction of stress when each individual is aware of how to safely handle food (Labrenz, M. 2022)

- Promotes communication.

It is in rare occasions that you will find employees willing to share potential food safety hazards with their managers. With continuous training, employees are able to get a sufficient platform to be able to raise their issues and concerns regarding the same. This is because at this platform, employees are always encouraged to provide their feedback. Comfortability in communicating with your manager regarding food safety and the potential hazards goes hand in hand with increased safety, a happy and productive environment. It should be noted that even when training and education is not considered to be a legal requirement, it is an aspect worth investing in as it will be able to boost the company to a better position.

Supporting research on food safety and contamination

The research on food safety and contamination is aimed at improving products' safety, as well as changing the perception of food products in society. Conducting research in a specific area is only the beginning. The next step is to use this knowledge to enable change, whether through policy or practice. Food safety and contamination policies can encourage farmers to operate in an ethical manner, but it is ultimately up to consumers to ensure that their food is safely prepared and properly handled. with support from the federal government and private industries, food safety and contamination research has dramatically improved over time. The United States Department of Agriculture (USDA) has funded research at universities nationwide in order to further knowledge about improving food quality. Apart from supporting universities worldwide research, the USDA also participates in the same with the overall aim of ensuring that the quality of food is maintained. In December 2012, USDA was able to implement the test and hold policy which has proven to be helpful regarding the reduction of cases of foodborne illnesses. Since the implementation, facilities tasked with researching a particular product in the market are required to hold it until the microbiological testing has already been done and determines it is safe to be released to the public regarding it as safe for human consumption. It was unfortunate that this policy came in too late as it would have prevented the 44 recalls of unsafe food that were experienced between the years 2007-2009.

USDA regards prevention to be the anchor when dealing with food products. As such it has focused on science, branding it as the basis for every decision made regarding food safety. It has been able to fund FSIS Agriculture Research Service (ARS), and in other agencies to conduct tests and analyses and further rely on them to inform their approach to food safety. All the data, science and research are the primary tools used to inform the same.

To have proper information regarding food, the USDA with other agencies is also researching various aspects of its safety. Food safety and contamination is a highly complex process. Food production involves hazardous practices, such as the use of pathogens and toxins in the manufacturing process; making food from unclean sources (food processing); use of improper packaging that may allow poor quality products to be brought to market; and improper storage methods, like using the wrong heat settings for a longer time period. Despite these hazards, research shows that people have the upper hand when it comes to keeping their food safe. That's because people take certain precautions when planning meals at home and make conscious decisions about what they are eating or cooking at home. This has led to many consumers being vigilant about what they are getting into their mouths. Others have even contemplated to the production of their own food products. When it comes to food safety, the reality is that we all want our food to be safe. The second realization is that there are many people out there who are not aware of how dangerous some foods can be to your health if consumed in large amounts. The USDA continues to study foodborne illnesses and other contamination issues that affect the integrity of the food supply.

Food that's been handled and prepared in an unsafe manner can cause illness or even death. Research has shown that the food we eat is contaminated by many things from pesticides to chemicals and even viruses, so making sure that you don't put your family at risk is essential. Researching food safety contamination is important to ensure that the products and the people who consume them are safe. Researching food safety contaminants helps the public understand what health risks are associated with the consumption of those foods and helps companies better manage those risks. Scientific research is necessary in order to create a foundation for efficient food safety systems and practices. Food safety must be continuously monitored to ensure that any and all contaminants remain below the allowable limits. By testing food samples for chemical contamination, and microbiological contamination using advanced techniques such as SEM-EDX, we can provide valuable insight into the risks surrounding our daily consumption of food.

Not researching food safety contamination can have some disadvantages, for example: the food may be contaminated and cause illness or death among consumers. This is especially in the current technological world where there are a lot of contaminations. Food safety research has been hindered by the lack of funding and resources. There hasn't been much research on this topic because of the lack of interest in creating a new product, finding out what is in our food and testing it, or knowing what chemicals are used to enhance flavors. This is just plain old ignorance and the only thing that is opening our eyes is the fact that individuals perish due to food contaminations. Supporting research to identify the causes of food contamination and to develop methods to safely ensure that foods are free from dangerous pathogens, such as bacteria and other harmful microorganisms can be the next step in ensuring that globally individuals are consuming safe products.

The U.S. Department of Agriculture is supporting research to determine ways to prevent and reduce the likelihood of food contamination as much as possible in order to maintain high-quality food products. The agency uses its influence to influence the food industry to create a positive change in the way food is produced and processed. When this happens, it impacts countless numbers of Americans and individuals all around the world. The effect of supporting research on food safety contamination by the department of agriculture is that it has led to a decrease in health hazards. The impact of basic science businesses has helped us understand what is important when we are studying diseases within the body. However, because many researchers find it difficult to obtain funding, they have difficulties carrying out research as they do not have enough money to pursue their projects. This leads to little progress being made in finding new cures and treatments for many diseases such as cancerous tumors and even some types of diabetes. As a result of the support for research on food safety contamination by the United States Department of Agriculture, it was found that pathogens can only be removed from cooked foods stored properly, not raw foods. Today's standards for cuts and serving meats in restaurants were based on research done at USDA.

Chapter

7 MEASURES FOR PROTECTING AGRICULTURAL SOILS FROM CONTAMINATION

Contents

7.1 Introduction.. 128

7.2 Control Technologies for Non-Point Source Contaminants 130

7.3 Position on the Issue of Contaminated Farm Runoff........................... 132

7.4 In Situ Physical/Chemical Treatment Technologies............................ 141

7.5 Bioremediation of Agricultural Soils.. 142

7.1 INTRODUCTION

A combination of biological, climatic, geological, and topographical events over time results in the formation of soil, which is a heterogeneous mixture of organic material and weathered rocks. It plays a significant role in our ecosystem. India's rural households still employ roughly 70% of people in agriculture, placing a significant burden on the economy's reliance on the quantity and quality of agricultural output. However, using pesticides has detrimental effects because it causes chemicals and salts to build up in the soil and move up the food chain (lower to upper trophic levels). The bacterial biodiversity in soil and microorganisms has been greatly reduced, which hurts soil quality. In addition, improperly disposed of industrial or municipal wastes affect the quality of the soil (Fagnano, M. 2018).

It is challenging to comprehend the serious effects that soil pollution has on the ecosystem as well as to deal with its eradication. To satisfy the rising demand for agricultural products, environmentally friendly practices must be used. To lessen the negative effects, the following drastic measures must be put into place.

- Commercial Waste Sorting

Based on how easily they can be disposed off, waste from various industries may be divided into different types.

- Biodegradable: Textile and paper mill waste, food processing waste
- Non-biodegradable materials include trash from thermal power plants, industrial wastes like fly ash and slags, mud residues from the production of metals like aluminum, zinc, and copper, and gypsum byproducts from the fertilizer sector. These might be used as raw materials in other industries. For instance, steel mill fly ash and slag may be used to create cement.
- Dangerous: wastes that are explosive or inflammatory, reactive substances like metals from industries, medications, colors, pesticide remnants, and rubber goods. At facilities for factory waste disposal, some hazardous wastes may be disposed of via controlled incineration techniques or by burning lesser quantities in open bins.

Organic solvents that accumulate in the soil over time might include dangerous chemicals that can be removed using extraction and separation methods.

- Using less and recycling more

To reduce soil pollution, it is essential to follow the three Rs of sustainable development (Reduce, Reuse and Recycle). Implementing these at the residential and industrial levels based on the sort of pollution is crucial. Products like cotton and paper might take the place of plastic and other non-biodegradable ones. Potentially, plastics might be utilized to build motorways, landfills, and other buildings. Newer technologies can transform plastic into garments and sustainable energy. Electricity may be created from garbage by growing bacterial species capable of turning metals, plastic, glass, and paper in junk into manure and biogas (Abramov, Y., & Alborova, M. (2020).

- Methods of external soil Treatment

To eradicate volatile contaminants from soil, heat treatment processes could be employed. Chemical treatment with proper solutions can assist in soil detoxification. Microbial treatments in which select bacteria are lab harvested to digest certain compounds and remove polycyclic hydrocarbons or oil spills in soil.

- Reforestation

Planting additional trees, crop rotation, and mixed cropping can assist to protect soil from erosion and enhance soil fertility.

- Proper sewage system maintenance and disposal of solid waste formed in residential areas may replenish soil nutrients, making it cultivable.
- The presence of dangerous chemicals and contaminants in the soil at levels above the usual proportions causes an ecological imbalance. This is termed soil contamination.
- Excessive agricultural activities, industrial wastes and discharges that disrupt soil flora and fauna, mining operations that erode the soil, and urban infrastructure activities that result in solid waste disposal in the soil all contribute to soil pollution.
- The employment of pesticides and other dangerous chemicals has resulted in a drop in product production.
- Factories, vehicles, and other urban facilities create harmful gases or pollutants, which ultimately mix with soil and have significant health repercussions.
- Polluted soil from surface run-offs transfers toxins to neighboring water sources, rendering them hazardous for use.

- Pollution damages the soil quality, which reduces the soil's capacity to support a varied range of plants and animals. The biological mix is hampered, and other distinctive qualities like salinity, nitrogen fixation, and general fertility are substantially weakened.
- Industrial waste segregation, following the philosophy of reducing, reusing, and Recycling wherever practical, adopting soil treatment technologies, ensuring reforestation, and appropriate timely maintenance of sewage systems all contribute to reduced soil pollution.

Initially, erosion reduces the nutrient-rich productive top layer of soil, limiting the soil's production capability. Soil erosion is separated into two types: fast erosion and geological erosion. Geological erosion is a natural event that happens as a result of the continual weathering and disintegration of rocks, with the rate of erosion staying lower than the rate of soil formation.

In accelerated erosion, however, the rate of soil erosion passes a given threshold and becomes rapid. Slash-and-burn agriculture, overgrazing, deforestation, mining, and intense and inadequate agricultural techniques are all too responsible for fast soil erosion. This higher rate of soil erosion eliminates organic matter and plant nutrients from the rich topsoil, lowering crop output. As a result, natural resource conservation and management are crucial. Although soil erosion cannot be stopped, it must be managed to a level that minimizes its negative impact on agricultural productivity and sustainability.

The combination of heavy metals, volatile organic compounds, dioxins, oils and fats, and other contaminants impacts soil and groundwater contamination conditions. Efficient restoration needs full engineering technologies that involve soil surveys and analyses, countermeasure design, execution, monitoring, and follow-up care, such as the production of proposals for empty lot utilization (Fagnano, M. 2018).

7.2 CONTROL TECHNOLOGIES FOR NON-POINT SOURCE CONTAMINANTS

Non-point pollution is commonly used to refer to low-level contamination that spreads over a large region. The key mechanisms governing contaminant interactions under these conditions include leaching, surface runoff, and sorption-desorption processes, and plant uptake. Yet certain pollutants, especially organic ones, may be broken down by processes including volatilization, chemical breakdown, and biological degradation. The main processes that control the movement and fate of organic pollutants in soils are sorption and desorption, respectively, and degradation (both biotic and abiotic). These procedures are affected by the soil and solution conditions present. The bioavailability and/or transport of pollutants in soils are both influenced by these interactions. Increased bioavailability of pollutants increases the danger they pose to soil, crops, human health, and both surface and groundwater.

Age-related non-point source pollution has long been recognized as a major contributor to eutrophication. For instance, 52 and 54% of the total nitrogen (TN) and phosphorus (TP) loading, respectively, in the Taihu Lake Basin, China, are attributed to agricultural non-point source pollution. Comparable percentages may be seen in Italy, where they account for 24% and 71% respectively. Agricultural non-point source pollution is suspected to be the primary contributor to nutrient depletion in American lakes and streams. The correct management of agricultural runoff and animal waste is of utmost importance to the United States Environmental Protection Agency (USEPA) and the United States Department of Agriculture (USDA).

What is known as "agricultural runoff" is the surface runoff from farmland that occurs when there is an excess of water on the land. The majority of its surplus water comes from both irrigation and rain. Nitrates, ammonium, phosphorus compounds, heavy metals, and persistent organic contaminants are all found in agricultural runoff. When eutrophication occurs, nitrogen (N) and phosphorus (P) are essential because they are used to make amino acids and DNA, respectively, which are used to produce aquatic plants (Fagnano, M. 2018).

Anthropogenic eutrophication has emerged as the leading issue because of the widespread agreement that it has the potential to significantly impact the security and stability of aquatic ecosystems worldwide. The closing of a Wuxi drinking water facility owing to the large "cyanobacteria mat" in Taihu Lake caused a catastrophe that impacted millions of people. Many species are put in danger by the buildup of heavy metals and persistent organic pollutants (POPs) in agricultural runoffs (e.g., pollution of drinking water). Therefore, protecting lakes and rivers from eutrophication, protecting the quality of drinking water, and preserving the aquatic ecosystem all require lessening agricultural non-point source pollution. Pollution prevention from agricultural runoff can be accomplished in one of three ways: at the source, during processing, or thereafter. Conservation tillage, fertilization management, and efficient irrigation systems are all examples of source control practices that can prevent leaching Process control aims to remove contaminants by making use of the time and distance that agricultural runoffs travel from the farm to receiving water sources such as ecological ditches. They are typically buried in rural drainage systems. If the levels of contaminants in the receiving water do not decrease to an acceptable level, then the only option is to resort to final treatment. Longer time is needed for agricultural runoff to be treated thanks to the enormous storage capacity. Although the strategies differ in their underlying assumptions, they all share a common goal: minimizing the environmental damage done by farm runoff. It might be difficult to locate programs that effectively combine available forms of care. In this analysis, we focus on both established and emerging technologies. A scenario analysis was also performed using the reference data to offer a comprehensive knowledge of the present agricultural runoff control strategies and their roles in effective agricultural runoff management (Fagnano, M. 2018).

7.3 POSITION ON THE ISSUE OF CONTAMINATED FARM RUNOFF

An economy may expand and thrive thanks to agriculture and animal husbandry. Large-population countries like Indonesia, China, and India would tremendously benefit from this. Chemical pesticides and fertilizers have grown more and more required in recent years to fulfill the massive global food demand. Fertilizers containing phosphorus (P) and nitrogen (N) have historically been utilized extensively. The FAO estimates that in 2015, 68.6 kg of nitrogen fertilizer and 30.1 kg of phosphorus fertilizer were applied on average per hectare of agricultural land (kg). In the US, fertilizer usage is once again on the rise. The world's greatest producer and consumer of fertilizers are China.

Chemical fertilizer use causes several environmental problems, including eutrophication of surface waters, emissions of greenhouse gases from nitrogen (N) breakdown, and pollution of groundwater. After rain and irrigation, surface water runoff adds to the nitrogen and phosphorus that is applied as fertilizer by transferring more to the crops. The transfer of nitrogen and phosphorus promotes the environmental and temporal conditions necessary for optimum recovery while also increasing system complexity. Since dyes are the foundation of an irrigation system, they might also serve as the main channel for drainage for agricultural land. Ditch areas can be useful for managing nitrogen (N) and phosphorus (P), as agricultural runoff has a certain amount of time to move before discharging into receiving water.

The goal of in situ technology is to maintain the original location of the treated soil. Soil alteration to introduce cleaner-acting chemicals may be necessary for some in-situ treatments. The contaminated component can be gathered, treated, or disposed of elsewhere thanks to the processes that isolate or concentrate contaminants. The capacity to treat soil without digging and moving it is the core advantage of in situ treatment. This can result in significant financial savings and fewer harmful environmental effects. Soil structure, organic matter, and biodiversity are all easily lost after invasive treatments, but they may all be preserved with in-situ care. In contrast, in situ treatment may need longer periods owing to a different site and soil characteristics. The effectiveness of a treatment is more challenging to prove, and its reliability is less certain (Abramov, Y., & Alborova, M. 2020).

For ex situ activities, the need for dirt removal exists. The removed dirt can either be transported off-site or treated locally (and then utilized again). Ex situ procedures take less time and entail fewer steps, but excavation and transportation costs are higher. Due to the disturbance of soil structure and organic matter caused by these treatments, recovering ex situ treated regions can be more challenging, expensive, and time-consuming.

Onsite technologies based on organic methods organisms that live in the soil are used in nature-based soil remediation techniques to biodegrade, stabilize, or segregate pollutants. Plants, soil microorganisms, and microorganisms are all used in the processes (bacteria, fungi, and archaea). For instance, as part of a nature-based remediation method, the symbiotic relationship between bacteria and plant roots may aid in biodegradation. The range of organic and inorganic contaminants dealt with by nature-based methods is continuously expanding. The contamination of a substrate by petroleum hydrocarbons, chlorinated solvents, polycyclic aromatic hydrocarbons, pesticides, and trace elements has been successfully remedied using environmentally friendly methods. The amount of time required to get the desired degree of cleaning may differ greatly depending on the approach, concentration, and kind of pollutant. While wetland phytoremediation may take years to complete, certain types of ex situ bioremediation can be finished in as little as 12 weeks. Increased biological activity in soil may be achieved in several methods, hastening the restoration process. Numerous studies have shown that the use of tailored Technosols or dietary supplements in conjunction with natural cleaning techniques can speed up the biogeochemical processes involved in cleaning. Increased bioleaching and biovolatilization of Arsenic are two examples of these processes.

- Bioremediation

An in-situ biological treatment known as "bioremediation" uses soil microorganisms to break down organic pollutants such as solvents, insecticides, and petroleum hydrocarbons as well as to alter trace element species to reduce their availability. These biological treatments are appropriate for both aerobic and anaerobic settings. There are several potential treatment methods, each with a distinct set of operating conditions and a list of contaminants that may be eliminated. To stimulate aerobic microorganisms, the aerobic biological treatment uses periodic intake of ambient oxygen through the soil. But during anaerobic biological treatment, which is frequently aided by reducing medicines, oxygen is eliminated. Aerobic biological treatment is frequently used to remove non-chlorinated or weakly chlorinated hydrocarbons, but anaerobic biological treatment is generally employed to breakdown highly chlorinated hydrocarbons, even though both aerobic and anaerobic biological treatment may degrade certain contaminants by your research.

The ability of a microbial population to break down organic contaminants in soil is increased through bio-stimulation and bio-augmentation, respectively. This process is known as enhanced bioremediation. The breakdown of organic pollutants is encouraged by increasing the microbial habitat through aeration, nutrient addition, pH control, and temperature modulation. Manure and sewage sludge are examples of materials with high nutrient concentrations, labile carbon, and microorganisms that may be added to speed up biodegradation. Sludge from sewage treatment plants and various forms of manure needs to be treated carefully since they contain potentially dangerous organic and inorganic contaminants. However, introducing unfamiliar microorganisms into a soil biome that is different from their own can occasionally backfire, leading to unexpected changes in the way local ecosystems work and the diversity of animals that live there.

The use of genetic techniques has produced the successful discovery, improvement, and restoration of native bacteria with the ability to degrade organic pollutants. Variables including the kind and characteristics of the soil and the regional climate have an impact on the rate and effectiveness of in situ biological treatment (Fagnano, M. 2018).

Some pollutants such as the explosives RDX and HMX, highly chlorinated organics, and hexavalent chromium that may not be susceptible to biodegradation under aerobic conditions can be efficiently handled anaerobically. To exclude oxygen, the soils are often wetted, and organic and inorganic amendments are given to promote the anaerobic microbes. Electron donors can be given to the soil to establish reducing conditions that accelerate the reductive dechlorination of chlorinated organic molecules

The basic method of microbiological remediation of soil contaminated with trace elements is to immobilize them and lower their bioavailability. Trace elements cannot be destroyed by microbes, but they can be transformed into another form with less dangerous physical and chemical characteristics. For example, aerobic and anaerobic bacteria may convert hexavalent chromium to its trivalent form which is less hazardous and less mobile Biodegradation may be increased by combining both phytoremediation and microbiological bioremediation approaches.

- Bioventing and bio-sparging

Bioventing and bio-sparging are related processes that enhance the biodegradation of any aerobically degradable substances by the direct infusion of air into the soil. Bioventing is conducted in the soil vadose zone and bio-sparging in the saturated zone below the water table. Air is delivered at low rates adequate to activate and sustain existing soil microorganisms to biodegrade the organic pollutants. The low rate of air flow attempts to decrease the volatilization and release to the atmosphere of organics and is typically used to remediate soil contaminated with less volatile hydrocarbons. In the instance that there are volatile pollutants, it might be essential to implement vapor extraction and treatment such as active carbon filtration or catalytic oxidation. The procedures are commonly employed to remediate soil that is contaminated with less volatile petroleum hydrocarbons where there are fewer emissions of volatile vapors. Many governmental authorities advise on cleanup technology (Abramov, Y., & Alborova, M. 2020).

- Phytoremediation

Phytoremediation refers to in situ approaches that employ plants for treating contaminated soils. Phytoremediation performs two major functions in remediating contaminated soils: stabilizing the pollutants so that they are less mobile and less available to cause harm, and eliminating them by aiding their breakdown or transferring them to different media. Depending on the plant mechanism in the remediation process and contaminant type, the phytoremediation may be split into different techniques: phytostabilization, phytoextraction, phytodegradation, rhizodegradation, and phytovolatilization, which are explained in detail below.

Phytoremediation has been used widely for remediating soils that have been contaminated with trace elements but can also be useful in promoting the removal and biodegradation of organic pollutants. The success of phytoremediation procedures is greatly determined by the properties of the pollutants and soil, as well as the selected plant species.

It also depends on the interactions of the plant's root systems with the microorganisms in the soil its efficacy can be boosted by adding amendments to the soil, for example, chelating agents can be added to increase trace element mobility. Amendments that modify the soil pH might enhance or reduce the mobility of trace elements. Kinetics has also been utilized to promote the mobility and absorption of trace elements by plants Improving soil fertility can also aid phytoremediation.

The effectiveness of phytoremediation can be increased by adopting indigenous plant communities that are tolerant to the unique environmental circumstances of the site, such as the degree and kind of pollution and climatic conditions. However, the germination and establishment of vegetation directly on polluted soils, even employing tolerant plant species, might be problematic due to other severe properties of the polluted soils. In general, the growth of plant cover leads to a decrease in soil pollution, reduced soil erosion and runoff, and an enhancement of landscape and soil functions (Fagnano, M. 2018).

Phytoremediation processes may be boosted by altering plant growth traits and growing circumstances by enhancing soil conditioning and management procedures. Depending on the phytoremediation technique and toxins involved, augmentation of plant features might focus on the change of the above-ground biomass, for example increasing it for phytoextraction. The upgrades might also focus on changing the features of the root system (for example the shape, surface area, microbial interactions, and root exudates) (Abramov, Y., & Alborova, M. 2020).

- Phytostabilization

Phytostabilization employs contaminant-tolerant plants to consolidate and stabilize contaminated soils to prevent the spreading of pollutants by wind and water erosion. The evapotranspiration of the plants decreases the moisture content in the soil and the flow of leachate, therefore reducing the spread of the toxins farther into the earth and preserving groundwater.

The major purpose of this phytotechnology is to limit the availability and mobility of pollutants in the soil. The immobilization of trace elements in the soil relies on the plant type and site circumstances. It can entail the production of compounds with limited solubility, chelation, adsorption onto the surface of roots, and absorption by the roots. Plants' roots may influence the rhizosphere conditions, in particular pH levels, and exudate chemicals which can have an essential impact in influencing the availability of trace elements

- Phytoextraction

Phytoextraction techniques absorb both inorganic and organic pollutants from the soil by direct absorption into plants. The rate and quantity of the pollutants that are transmitted from the roots to the shoot system is depending on the contaminant and the plant species.

Unlike organic pollutants, which can serve as a source of carbon and energy for soil microbes, trace elements are not biodegradable and cannot be chemically eradicated. Instead, they can infiltrate living tissues, accumulate (bioaccumulation) and concentrate (biomagnification) in the food chain. Several trace elements play a crucial role in plant development and metabolism (iron, manganese, zinc, copper, magnesium, molybdenum, and nickel), whereas others do not have a recognized purpose (cadmium, chromium, cobalt, lead, or mercury). Certain plant species are documented hyperaccumulators of these non-essential trace elements, absorbing trace elements through their roots and translocating them to their shoots, contributing to the remediation of the contaminated soil. However, this translocation can also lead to the entrance of trace elements into the food chain when trace element-enriched parts are ingested by livestock, wild animals, and humans. a global database of hyperaccumulator plants, useful not only for research purposes but also for designing phytoremediation programs as well as for risk assessment

- Phytovolatilization

Phytovolatilization processes involve specialized plant enzymes that can transform and volatilize inorganic and organic contaminants in the plant–microorganism–soil system. In essence, the contaminant is transferred from the soil into the plant, where is transformed and released into the atmosphere. This technique has been used for soils polluted with inorganic and organic mercury, where the plants can reduce Hg2+ to elemental mercury, which is less toxic and is ultimately volatilized from the plant. As the contaminants are transferred from one medium to another, careful risk assessments need to be undertaken to ensure the appropriateness of the approach.

- Rhizodegradation

Rhizodegradation is the breakdown of organic contaminants in soils by fungal and microorganism activity associated with the root zone. This plant-assisted biodegradation is sustained by root exudates that can enhance microorganism activity and biodiversity, and support the biotransformation of organic contaminants through co-metabolism. Cultivation of *Phragmites australis* in dredged sediments was demonstrated as an exploitable biostimulation approach to accelerate the depletion of PAHs. *Phragmites australis* accelerated the oxidation of the PAHs by rhizodegradation (Abramov, Y., & Alborova, M. 2020).

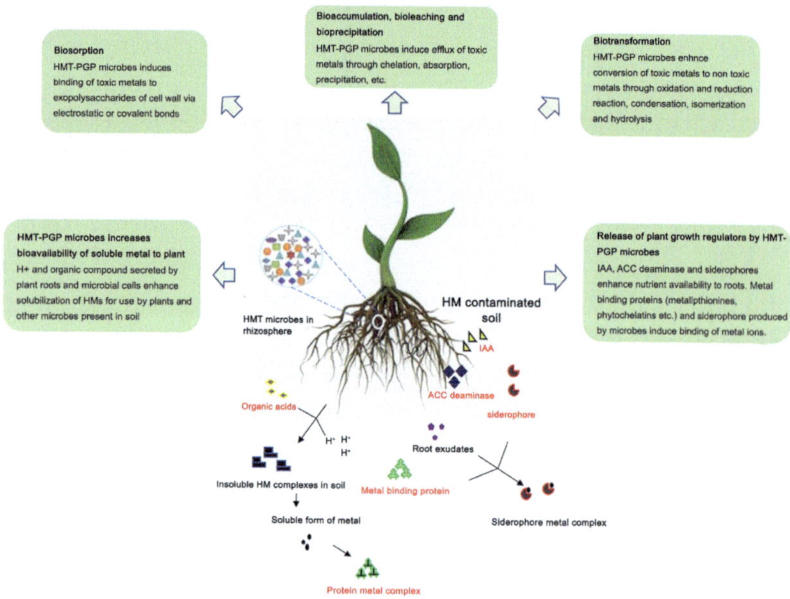

Figure 7.1 Phytoremediation.

Source: Phurailatpam L, Dalal VK, Singh N and Mishra S (2022) Heavy Metal Stress Alleviation Through Omics Analysis of Soil and Plant Microbiome. Front. Sustain. Food Syst. 5:817932. doi: 10.3389/fsufs.2021.817932

- Vermiremediation

Although many contaminants are toxic to earthworms, and they are often used as bio-indicators for polluted soils, under suitable conditions they can be used in remediation strategies. Earthworms can be tolerant to many chemical contaminants including trace elements and organic contaminants in soil and can bio-accumulate them in their tissues. Earthworm species have been found to remove trace elements, pesticides, and lipophilic organic contaminants, including polycyclic aromatic hydrocarbons (PAH), from the soil. They absorb the chemicals from the soluble soil fraction in the interstitial water through their moist body wall and also ingest them. As the soil passes through the earthworm gut contaminants are either bio-transformed or biodegraded rendering them non-toxic.

Meanwhile, the quality of the soil is improved significantly in terms of physical, chemical, and biological properties as the worms thoroughly upturn and disperse the soil, ingest large volumes of soil and excrete nutritive materials (nitrogen, phosphorus, potassium, and micronutrients) in the form of vermicasts along with millions of beneficial soil microorganisms including nitrogen fixers. It was demonstrated that in soil polluted with 20 g/kg - 60 g/kg of petroleum, the content of hydrocarbons decreased by 99 percent after 22 weeks in the presence of worms and a proprietary mix of nitrogen-fixing bacteria, yeasts, and fungi

- Ex situ biological treatment

Ex situ biological treatment requires the soil to be excavated and treated on-site or at a specialist soil treatment facility. The bioremediation processes are similar to those used in situ techniques. However, ex-situ techniques tend to allow greater control of the bioremediation processes through easier monitoring, control over the homogeneity of the soil, more effective mixing and contact between contaminants and microorganisms, and the effective application of nutrients and amendments. Ex situ processes tend to be faster and more predictable but more expensive than they're on the same lines as in situ ones. Ex situ processes tend to be more disruptive to the soil structure and it is biological communities, and post-remediation rehabilitation can take longer (Fagnano, M. 2018).

- Biopiles

Biopiling is a bioremediation technology, extensively used for ex situ remediation of polluted soils by a wide range of petrochemical and other organic chemical contaminants. The technique involves the excavated soil being homogenized and amended to enhance the conditions for biodegradation by microorganisms. The piles are often constructed and manipulated in the same way as a compost windrow. Biopiles are sometimes referred to as biocells. The activity of microorganism populations, growing near optimum conditions, can be stimulated aerobically or by the addition of nutrients, minerals, and moisture or pH adjustments to maintain optimum biodegradation

- Bioleaching

Bioleaching is an extractive technology that uses a solution inoculated with microorganisms to leach trace elements from polluted soils. It is also a process that is used in the mining industry as an alternative to cyanide to extract metals from low-grade ores or mine wastes. It is considered an efficient and cost-effective alternative to physicochemical treatment technologies for soil remediation. The inoculated liquid is trickled over an ex-situ pile of polluted soil. As the liquid percolates through the soil, the microorganisms act on the metal contaminants, which dissolve into the leachate and collect at the bottom of the pile. The leachate is further processed to recover or safely dispose of the metals. A potential weakness of the technology is the potential for flow channels to develop, with the bypassed soil remaining untreated.

- Composting

Composting is an aerobic ex situ technology, similar to biopiling, which involves mixing the excavated polluted soil with the non-hazardous bulking agent and organic amendments (e.g. straw or green waste) to provide the optimum levels of air and moisture to microorganisms. Generally, the mixture contains 75 percent of polluted soil and 25 percent of organic fraction but may be varied to suit the type of soil, contaminants, and concentration. To stimulate the composting process, the pile may be mechanically churned to homogenize the mixture, maintain aeration, and ensure even biological activity throughout the pile

- Bioreactors

Bioreactors are open or closed installations that agitate the soil while adding nutrients to stimulate microorganisms to degrade the contaminants. They may be operated aerobically or anaerobically depending on the contaminant and the microorganisms best suited for biodegradation. Typically, the time required for remediation is significantly shorter than static remediation strategies. There are two main types, solid phase, and liquid phase. A solid phase bioreactor is generally a batch process undertaken in an enclosed reaction chamber with internal stirrers. The process proved effective in treating tropical soil polluted with up to 35 percent total petroleum hydrocarbons, over 42 days. In a liquid-phase bioreactor, polluted soil is sprinkled with water and agitated with mixers. It is also a batch process. When the desired level of contamination has been reached, the soil solids can be filtered and returned to the site and the water retained for the next batch. The biodegradation process in both solid and liquid phase bioreactors is highly controllable and generally rapid due to the homogeneity of the materials and the effective contact of the microorganisms with the contaminants (Fagnano, M. 2018).

7.4 IN SITU PHYSICAL/CHEMICAL TREATMENT TECHNOLOGIES

Physical and chemical treatment techniques tend to be more aggressive than nature-based processes but are capable of treating soils with more persistent contaminants or those containing higher contaminant concentrations. The processes tend to achieve the remediation objectives more rapidly than nature-based equivalents. However, they tend to be costlier, and resource intensive, and, following treatment, it is often more difficult to re-establish biodiversity and soil quality in the treated soils.

Electrokinetic separation

Electrokinetic remediation uses electrical fields and electrochemical processes to enhance the migration of polar inorganic and organic molecules out of the soil. It is useful to increase the extraction rates for contaminants in low-permeability soils. It has been used in conjunction with in situ phytoextraction in Guiyu China as shown in the case study below. The technique was also used to remove trace elements from polluted soil in the Janghang smelter case study from the Republic of Korea (Abramov, Y., & Alborova, M. 2020).

Soil vapor extraction and air sparging

Soil vapor extraction and air sparging are similar processes to their biological equivalents, bioventing, and bio-sparging but rely solely on the air (which may be heated) passing through the polluted soil to volatilize the volatile and semi-volatile contaminants. These processes are most effective in soils with high porosity. They are not effective in clay soils. Soil vapor extraction is the term for the process operated in the unsaturated vadose layer. Unlike bioventing, it relies on extraction pumps to provide a vacuum that draws air through the soil. Passive vents may be sunk into the soil to allow air to be introduced into the polluted zone. In air sparging, the air is injected below the groundwater level. The contaminants transfer from the soil and groundwater to the air bubbles, which migrate upwards into the vadose zone. As the objective of both these processes is to volatilize the contaminants, the airflow rates tend to be high. Although the focus is on extraction, the high air flows can facilitate aerobic biodegradation of the contaminants. In both processes, the air and volatilized contaminants are captured by the vacuum extraction system followed by a treatment process for the volatile. The treatment system would be specific for the flow rate of the extracted air and the type and concentration of the contaminants. For solvents and petroleum hydrocarbons, typical treatments include recovery, catalytic oxidation, and active carbon filters.

7.5 BIOREMEDIATION OF AGRICULTURAL SOILS

Remediation is the process of eliminating a problem, and it is known as bio-remediation when it involves eliminating an ecological problem, such as soil and groundwater contamination. A technique called bioremediation uses active microorganisms to lessen or stop natural pollution.

It is a step-by-step process that will eventually rid the atmosphere of toxins, restore the basic elements of the ecosystem, and stop future pollution. Bioremediation can offer a long-term in-situ remedy to pollution rather than just moving the issue. Heavy metals, metalloids, and other inorganic contaminants can be eliminated from soil or water with this method. It is a technology with the high public acceptance that is cost-effective, effective, unique, eco-friendly, and powered by solar energy as compared to other engineering methods (Abramov, Y., & Alborova, M. 2020).

The two groups of bioremediation approaches can be separated based on the methodologies employed.

- In situ bioremediation.

Ex-situ, as the name suggests, refers to transporting pollution to a distant treatment facility. Because it necessitates the laborious process of extracting contaminated soil and moving it away, this categorization is not commonly utilized. Ex situ remediation's guiding philosophy is to create ideal soil oxygen, moisture, and nutrient conditions away from the site. Ex situ bioremediation, on the other hand, runs the danger of accidentally spilling anything during shipment. There are two technique classes available, which are described in more detail below.

- Phase Slurry

To maintain the naturally occurring microorganisms in touch with the toxins, this approach involves mixing contaminated soil with water and other additives in a large bioreactor. In the bioreactor, vital nutrients and oxygen are provided, and the environment is ideal for microorganisms to break down pollutants. After the treatment is complete, the water is separated from the particles; if the wastewater is still polluted, it is disposed of and subjected to further treatment. When compared to other biological treatment methods for polluted clays, slurry-phase treatment is a comparatively short procedure.

- Solid-state

Solid phase treatment is used to remediate the soils in the above-ground treatment area. Collection systems are installed here to prevent impurities from evading treatment. The pace of deterioration is accelerated by controlling the quantities of moisture, heat, nutrients, and oxygen. Solid-phase systems demand more space and treatment time than slurry-phase procedures, even though they are simpler to process and maintain. To obtain this therapy, the approaches listed below can be applied.

- The agricultural sector

In essence, this method encourages biodegradation using local microbes and permits the aerobic decomposition of pollutants. It is done using a simple approach procedure wherein contaminated soil is removed and put over a ready bed regularly until contaminants are broken down. The development of native species is also encouraged by the addition of some fertilizers and minerals.

- Soil piles

Petroleum-contaminated excavated soil is remedied using this biodegradation method. Biocells are another name for soil biopiles. Using this approach, polluted soil is piled up while microbial activity is encouraged either aerobically or by the addition of nutrients, minerals, or moisture. Biopiles can be anywhere between three and 10 feet tall. Additionally, oxygen is used in this method to promote bacterial growth. By forcing air to circulate via perforated tubes positioned throughout the pile, aerated biopiles are produced

- Composting

To provide the microbes with the right amount of air and water, contaminated soil is mixed with biomass, such as straw, hay, or corncobs. This process is known as composting. Contaminated soil is mixed for aeration in treatment containers before composting. In a specific kind of composting called window composting, the earth is piled up in very long mounds called windows and continually churned by tractors. A 75% contaminated soil to 25% compost ratio is utilized for composting. This ratio changes depending on the kind of soil, the number of contaminants, and other factors. Because compost remediation may be finished in a few weeks, it is seen as a quicker remediation technique.

In-situ bioremediation

The bioremediation procedure carried out at the polluted site establishes the in-situ technique. Because it needs less mechanical work to remove spreading pollutants and stop pollutant spread through transportation or pumping away to other treatment locations, in situ bioremediation is the recommended option. The term "in situ bioremediation" describes biological procedures in which microorganisms change organic pollutants into inorganic substances such as carbon dioxide, methane, water, and inorganic salts. Both natural and artificial environments can be used for this method.

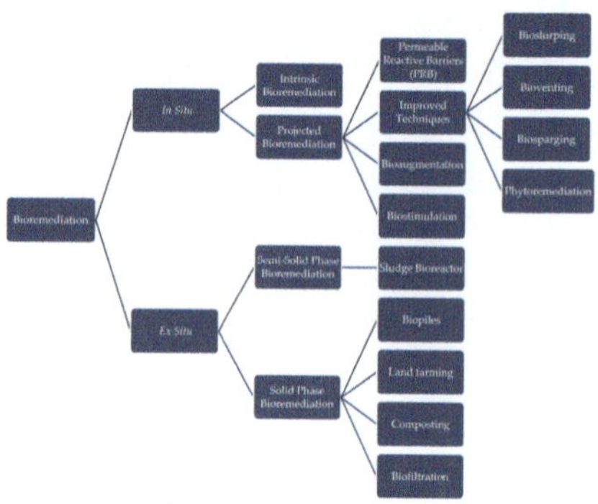

Figure 7.2. Bioremediation.

Source: Sales da Silva IG, Gomes de Almeida FC, Padilha da Rocha e Silva NM, Casazza AA, Converti A, Asfora Sarubbo L. Soil Bioremediation: Overview of Technologies and Trends. Energies. 2020; 13(18):4664. https://doi.org/10.3390/en13184664

Various Bioremediation Methods

- Bioremediation from inside

Intrinsic bioremediation is a procedure that transforms environmental contaminants into non-toxic forms by utilizing the intrinsic capabilities of the site's naturally existing microbial community. This method is frequently applied in subterranean spaces, such as petroleum storage tanks. Intrinsic bioremediation uses naturally occurring microbial communities' intrinsic capacity to break down environmental contaminants without altering or speeding up the process. To boost the metabolic activity of the local microbial community, this strategy entails providing it with nutrients and oxygen.

- Better (engineered) In Situ Bioremediation

This procedure involves introducing certain microbes to the polluted place, as the name would imply. By enhancing the physical, chemical, and biological conditions for microbe growth, engineered in situ bioremediation quickens the degradation process.

- Bio-ventilation

An in-situ remediation method called bio-venting uses microorganisms to break down organic components that have accumulated on soils. By increasing the amount of oxygen reaching the unsaturated zone, this approach increases bioremediation by boosting the activity of local bacteria. To boost bioremediation and accomplish the microbial transformation of contaminants into a harmless state, the bio-venting process is modified by adding nutrients and moisture.

In comparison to other in situ bioremediation methods, this one has gained popularity, especially for the restoration of areas that have been contaminated by minor petroleum spills. However, it may also be utilized to degrade volatile organic compounds (VOCs) via biologically active soil, which is how bioventing is predominantly employed to decompose adsorbed fuel residuals.

Bioslurping The bioslurping method combines bioventing with vacuum-enhanced pumping to remediate soil and groundwater by indirectly supplying oxygen and stimulating pollutant biodegradation Using a "slurp" that extends into the free product layer, this method sucks liquids (free products and soil gas) from this layer like how a straw draws liquid from a vessel. A vacuum pump is erected and attached to a well with an adjustable length referred to as a "slurp tube."

This vacuum pump is lowered into the light non-aqueous phase liquids (LNAPL) layer and pumping begins to extract free product as well as some groundwater. The LNAPL flow is facilitated by the vacuum-induced negative pressure zone in the well, which also pulls LNAPL trapped in tiny pore gaps above the water table. This technique is used to remove organic molecules that are semi-volatile and volatile from the soil (Abramov, Y., &

Alborova, M. 2020).

- Bio-sparging

A biological technique called "bio-sparging" can be used to clean up an area of aromatic substances including benzene, toluene, ethylbenzene, xylene, and naphthalene. Specific aerobic microorganisms are used to convert aromatic and mineral compounds into simpler and more usable forms. Similar to bioventing, this approach involves injecting air into the soil's subsurface to promote microbial activity and enhance the removal of pollutants from contaminated regions. By injecting air into the saturated zone during bio-sparging, volatile organic molecules ascend to the unsaturated zone and encourage biodegradation. There are two main influences on the bio-sparging process:

- Soil permeability (which determines pollutant bioavailability to microorganisms)
- Pollutants' ability to biodegrade
- Bio adjustment

Enhancing the current microbe population to increase its capacity for lowering pollution levels is known as bioaugmentation. Using this method, polluted soil is mixed with organic culture to produce a setting akin to a bioreactor. There are two typical options: adding a pre-adapted consortium to the polluted location and adding a pure bacterial strain. At oil-contaminated locations, bioremediation is the main use of bioaugmentation. Bioaugmentation is a low-cost approach to wastewater and soil pollution treatment compared to other techniques.

In conclusion, one effective method for decontaminating polluted environments is bioremediation. Bioremediation is not a novel idea. On the other hand, some uses are relatively new, and many more are developing or emerging. This process can be either aerobic or anaerobic, depending on the bacteria and electron acceptors present. Intrinsic bioremediation is one option for this process, whereas human intervention is another (engineered bioremediation). The distinction between soil and groundwater has limited practical importance in this context since many remediation processes, particularly physical systems, require the treatment of aqueous phase pollutants. A quick examination of remediation methods that are just intended to treat or confine groundwater inside "geological" materials will be followed by a more in-depth discussion of methods that are frequently used for both treatment and containment. These techniques offer an efficient and affordable way to clean up contaminated soil and groundwater. Reforestation and waste material recycling are two more common methods of preventing soil pollution. Because soil fertility is lost as a result of deforestation, soil erosion usually results in soil pollution. Soil pollution can be decreased effectively by replanting. Another effective and frequently used method of reducing soil contamination is recycling items like plastics, papers, and a variety of other materials to reduce the amount of trash or waste that ends up in landfills.

Overall, research has found that pollution threatens human health, degrades the ecology, and limits our ability to grow food by destroying soils. Future generations can benefit from clean water, clean air, and healthy soils thanks to bioremediation's ability to reduce and remove pollutants. The bioremediation process is entirely natural and carries very few risks. The majority of applications that do not need dangerous transportation are completed on-site. There are seldom any dangerous byproducts produced. Since it requires far less equipment and labor than other cleanup techniques, bioremediation is significantly less costly.

Chapter

8

CHALLENGES TO THE PREVENTION OF AGRICULTURAL POLLUTION

Contents

8.1 Introduction....................150
8.2 Challenges to the Prevention of Agricultural Pollution....................151

8.1 INTRODUCTION

Agriculture consumes 70% of all freshwater used globally, making it a major contributor to water contamination. Because of the large amounts of agrochemicals, organic matter, drug remnants, sediments, and saline drainage released into waterways, farms become a major source of water pollution. Agricultural practices can have a significant negative impact on water resources, however, these impacts can be mitigated with careful management that includes diagnosis, forecasting, and monitoring.

According to a report published by the Food and Agriculture Organization of the United Nations (FAO) and the Water, Land, and Ecosystems (WLE) programme led by the International Water Management Institute, rising demand for foods with large environmental footprints, such as meat from industrial farms, is causing unsustainable agricultural intensification and a decline in water quality.

This increase in crop yield is primarily due to the widespread use of pesticides and chemical fertilizers. Pesticides have a market value of more than $35 billion per year. Pesticide use has risen by double digits in some countries like Argentina, Malaysia, South Africa, and Pakistan. (Kraemer, R. A., Choudhury, K., & Kampa, E. 2001)

According to FAO's Land and Water Division Director Eduardo Mansur, agricultural pollution has surpassed contamination from settlements and industries as the primary factor in the degradation of inland and coastal waters in most high-income countries and many emerging economies. Recognizing that a problem exists is the first step toward resolving it.

Nitrate is the most common chemical contaminant in groundwater aquifers throughout the world today. Agricultural pollution has an impact on aquatic ecosystems, for instance, eutrophication caused by nutrient accumulation in lakes and coastal waters affects biodiversity and fisheries. Despite the lack of data, 415 coastal areas were identified as eutrophic.

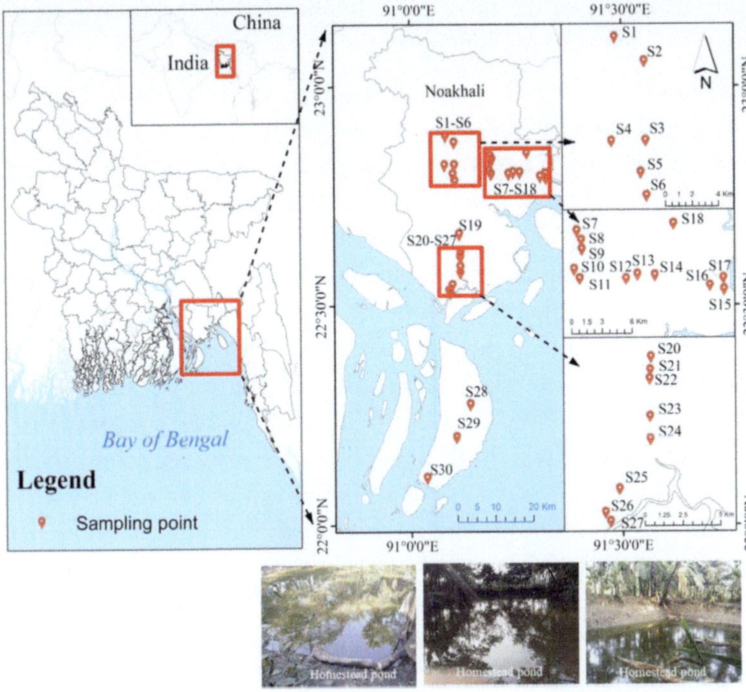

Figure 8.1. Toxic algal blooms.

Source: Akter L, Ullah MA, Hossain MB, Karmaker AR, Hossain MS, Albeshr MF, Arai T. Diversity and Assemblage of Harmful Algae in Homestead Fish Ponds in a Tropical Coastal Area. Biology. 2022; 11(9):1335. https://doi.org/10.3390/biology11091335There are two main types of agricultural pollution namely:

- Animal agriculture pollution
- Pollution from growing crops

8.2 CHALLENGES TO THE PREVENTION OF AGRICULTURAL POLLUTION

8.2.1 Animal waste disposal challenge

Meat was once considered a luxurious food. However, as individual wealth increased over the last century, so did our desire for the consumption of meat, dairy, and eggs. This has, in turn, led to an increase in the rearing of the associated animals to produce eggs, dairy and meat to cater to the increasing demands of the products. This has led to the following types of challenges in a quest to prevent agricultural pollution.

All animals do release a lot of waste in their lifetime. This waste is not treated and released like human waste. Instead, this waste is spread untreated on land for disposal. Operators are only supposed to apply as much manure as the crops can utilize, but this rarely happens because there is too much manure. As a result, more manure is applied than the soil can naturally absorb, which causes runoff into water sources. To make matters worse, the manure typically sits on-site in sizable manure lagoons before being applied to land. The lagoons are filled with a poisonous mixture of waste-decomposing bacteria, chemicals, and antibiotic residue that can take on a sickly hue. They frequently lack linings and are prone to spills, overflows, and leaks, which allows the contents to seep into the surface and ground waters. (Hodge, I. 2003). Similar issues can occur with poultry waste, which is primarily dry litter. A mixture of the birds' bedding materials (like shavings), their feces, and loose feathers that are kept in open and enormous mounds find their way into water resources. Chicken manure contains more phosphorus than other animal manures and therefore more toxic to the waterways.

Case study on manure management policies and practices in sub-Saharan Africa

Since ancient times, manure has been utilized as fertilizer. If properly handled, it can be a resource for sustainable agriculture and increased crop output, especially for smallholder farmers in sub-Saharan Africa (SSA). However, the majority of farmers in SSA do not implement advised manure management practices, such as covering manure during storage, having a water-proof floor, or roofing animal housing. This results in significant nutrient losses during manure storage increased greenhouse gas emissions and decreased manure quality as a fertilizer (Kraemer, R. A., Choudhury, K., & Kampa, E. 2001)

Improved methods for collecting, treating, storing, and applying manure to soils are the main components of integrated manure management (IMM). These methods can increase yields while also reducing nitrate (NO3) and phosphorus (P) leaching, ammonia (NH3) volatilization, nitrous oxide (N2O), and methane (CH4) emissions. Poor manure management has the potential to endanger public health, especially in cases where livestock is kept in urban and semi-urban settings. It can also result in the transmission of zoonotic illnesses to people. The GlobalGAP limits the application of manure to 60 days before harvest for other crops and prohibits applying untreated manure to leafy vegetables once they have been planted due to public health concerns. Poorly released manure can cause ground and surface water to become eutrophic and contaminated, primarily with nitrate. High nitrate levels in drinking water are widespread in southern Africa and harm people's health, especially children and the elderly.

Both in Ethiopia and Malawi, urine storage is rare. This is probably because the most prevalent confinement systems don't have an impermeable floor, which means that pee tends to seep into the ground beneath the kraal. This implies that a significant number of nutrients are wasted, particularly nitrogen and potassium, which could be harmful to the environment around them. Manure is typically kept in both nations as a solid. Over two-thirds of farmers in Malawi saved solid manure for subsequent use as fertilizer, but practically all farms in Ethiopia stored and dried solid manure that was later used for fuel. Smaller and medium-sized farms in both countries were more likely than bigger farms to employ manure as fertilizer because smaller and medium-sized farms also tended to contain cropland, whereas large farms were primarily focused on animals with little to no cropland. Additionally, none of the farms in Ethiopia and Malawi that were subjected to the interviews had roofing or covers for the storage of manure, exposing it to further nutritional losses.

Most of the manure produced by scavenging animals, especially the hens, was left lying around the house and was not typically removed. Farmers in Ethiopia's Awassa region frequently went on transhumance for up to three months with their livestock, which included small ruminants, while leaving the rest of the family behind. As a result, it is not possible to employ biodigesters in this situation because they need to be fed with fresh manure every day. In addition, the migration's waste, which acts as a fertilizer, will be dumped straight on the pasture lands.

Poor manure management in Ethiopia is frequently caused by a lack of farm labor, especially when the manure is liquid and must be transported carefully. However, some farmers combined agricultural waste and other materials with liquid manure to create compost, which is typically simpler to transport. Technical problems (such as a lack of tools, storage space, infrastructure for trading, etc.) were not regarded as obstacles to better manure management. In Ethiopia, one-third of farmers were satisfied with the return on their investments after spending money on better manure management within the previous five years (mainly higher crop yields). These farmers stated that rather than agricultural extension specialists, other farmers persuaded them to make these expenditures. Although information sharing among farmers was restricted to specific geographic areas, this reduced their ability to learn from one another and lowered the rate at which recommended manure management strategies were adopted.

Manure management has not been fully covered in the agricultural extension system's training program, and there is currently a procedural barrier preventing extension agents from advising farmers on better manure management. Part of how extension agents are evaluated and promoted depends on how much synthetic fertilizer they sell. Extension personnel don't usually encourage field use of manure because farmers tend to buy less synthetic fertilizer when they utilize more manure. The extension system should review its criterion for paying extension personnel, maybe basing it on an increase in crop yield per hectare (Kraemer, R. A., Choudhury, K., & Kampa, E. 2001)

The majority of farmers in Malawi said that lack of information was the main barrier to improved manure management. They lacked a sufficient understanding of manure management techniques and their potential advantages. However, farmers acknowledged the value of manure as a fertilizer and the majority felt it was essential for them. The majority of them noted that manure allows them to buy less fertilizer because synthetic fertilizers are expensive. Similar to Ethiopia, around a third of Malawi's large-scale farmers had spent time or money over the previous five years enhancing manure management. It would be wise for these farms to prepare compost manure since it is more marketable and convenient to transport than fresh dung because they often do not own cropland. Vast-scale crop farmers might purchase large quantities of manure from commercial compost producers, treat it, and then sell it to them. (Hodge, I. 2003).

Disposing of large sizes of waste from these animals is a great challenge which cannot be easily attained and this poses a major problem to any efforts towards the prevention of agricultural pollution to the water resources by animal waste.

Increased population growth leads to increased demand for food products. The population on the planet has doubled over the last decade leading to increased demand for food products. The increased demand has in turn led to increased agricultural activities to feed the increased population. Although food production in developing nations has risen significantly in recent years, it has only just kept up with population growth since the 1970s, and it hasn't done so in many low-income nations including Bangladesh, Nepal, and twenty-seven of the thirty-nine countries in sub-Saharan Africa. Only a minor rise in food production per capita has been achieved in some African nations, such as Kenya, Malawi, Rwanda, and Upper Volta.

Since the middle of the 1960s, food production in China and India has likewise outpaced population increase, albeit by a small margin. Increases in food production in the past were mostly brought about by cultivating more land; this is still the case in sub-Saharan Africa and some regions of Latin America. 3.4 billion hectares, or around 25% of the world's land, are believed to have agricultural potential. There is minimal evidence of worldwide land scarcity because only 1.4 billion hectares (or 40%) of this total are being used for agriculture.

The growth in agricultural productivity during the previous 20 years in emerging countries as a whole, however, was not entirely due to increasing land. This is due in part to the fact that land reclamation is frequently more expensive than intensifying the use of already-existing land and in part to the fact that many regions of the world are preventing further extension of the land border. For instance, in Sub-Saharan Africa, illnesses like river blindness (onchocerciasis) and sleeping sickness prevent the growth of large areas. The latter makes raising livestock in the roughly 10 million square kilometers of higher rainfall regions or 45% of all the land in sub-Saharan Africa virtually unfeasible. There have been significant initiatives to rid the Sudanese savanna country of sleeping sickness, but it hasn't always been feasible to stop the comeback of the illness. The ecology has also been negatively impacted by insecticides used to combat tsetse flies, which are the carriers of sleeping sickness. Some African nations are approaching their land's boundaries due to these and other factors. (Kraemer, R. A., Choudhury, K., &Kampa, E. 2001)

Increased population is a factor which has made it hard for those advocating for change in farming methods since the input might reduce when the farmers cease to use chemical fertilizers.

8.2.2 Ignorance by the farmers

Aquaculture, livestock, and crop production on-farm practices are essential for preventing pollution. Limiting and optimizing the type, amount, and timing of crop applications are two management strategies for lowering the risk of water pollution caused by organic and inorganic fertilizers and pesticides. It has been demonstrated that establishing protection zones alongside surface watercourses, inside farms, and buffer zones surrounding farms can effectively reduce pollution migration to water bodies. Additionally, effective irrigation plans will lower water return flows, which will significantly lower the migration of pesticides and fertilizers into water bodies (Hodge, I. 2003).

Figure 8.2. Farmers and fishers along the Mekong River and its tributaries use recession agriculture, a farming system of planting crops on fertile banks as flood waters recede to take advantage of soil moisture

Source: Morton, Lois & Olson, K.. (2018). The Pulses of the Mekong River Basin: Rivers and the Livelihoods of Farmers and Fishers. Journal of Environmental Protection. 09. 431-459. 10.4236/jep.2018.94027.

Some farmers are ignorant to use these methods. They will therefore apply the fertilizers regarding no guidelines which certainly will end up in the water resources. Preventing pollution when there are such people becomes a major problem which makes it hard to achieve any success.

It is impossible to overstate the importance of managing agricultural solid wastes. Solid agricultural wastes come from a variety of sources. Pesticides, which include insecticides and herbicides, are one of these sources. If all pesticide use were to cease, it has been calculated that the amount of food produced on a global scale would decrease by approximately 42%. All hands must be on deck to combat the threat posed by improper management of agricultural solid wastes because of the significant impact that agricultural solid wastes have on human health, animal health, and the environment. A large part of the improper management of agricultural solid waste is ignorance. Many of the farmers and household managers who produce these wastes are unable to handle them properly. Many of them are 'handicapped' because they are unaware of the health risks associated with the things they play with. Large amounts of agricultural solid waste are produced annually, increasing by about 7.5% each year. In many areas of developing nations, agricultural solid wastes are carelessly disposed of or burned in public spaces, causing air pollution, soil contamination, the release of hazardous gases, smoke, and dust, and the possibility that the residue will end up in a water source, polluting the water and aquatic environment.

8.2.3 Poor policy and structural factors

Farm pollution has increased as a result of policy and structural factors. A number of the pollution issues that countries are dealing with have been made worse by structural growth patterns, some of which are unique to the area. Development strategies that view the environment as a resource to be exploited have been encouraged by demographic increase and a society that places a strong emphasis on output and yield, influenced in part by the policy. Farmers generally believe that adding more fertilizer always results in higher yields, thanks to direct and indirect fertilizer subsidies, preferential input loans, extension messages, and product advertising. These elements have influenced the excessive usage of fertilizer. Specialization in crop agriculture, simplification of agroecosystems, and the transition to monoculture favorable to investments in irrigation infrastructure and extension messages, as well as occasionally rigid land-use policies have sped up the loss of natural predators and fertility on farms, increased the susceptibility of crops to pests and disease, and driven farmers' greater reliance on synthetic chemicals to address these issues.

The development of irrigated agriculture, a focal point of public agriculture investment in various nations, has unintentionally increased soil salinization, pesticide runoff, and rice-related greenhouse gas emissions. A more damaging mix of goods and activities, as well as closer proximity of inhabitants to agricultural sources of pollution, are the results of the agricultural sector's high degree of responsiveness to the rising demands of urbanizing populations. As a result of the increase in demand for animal products, animal agriculture has grown, including the expansion of restricted fish and pig farms. This has resulted in increasing pollution levels. Human exposure to the constant concentrating collection of pollutants produced by the agricultural sector has increased due to the concentration of agricultural activities and their increasing proximity to crowded population centers. Concerns about plastic, nutrient, and pesticide contamination have most recently arisen as a result of Chinese agriculture's response to the growing demand for fruits and vegetables in a water-scarce climate. A significant drawback of the quick development and expansion of agricultural production to date has been pollution.

The use of pesticides and other chemicals in farming has increased so quickly in a region that is committed to leapfrogging as a development strategy that it has occasionally outpaced farmers' and regulators' awareness of their dangers, the latter's capacity to regulate them, and the adoption of safe handling techniques. Creating a high-output seafood business quickly through the importation of aquaculture species designed for intensive cultivation has increased the danger of genetic contamination, disease, and biodiversity loss. Additionally, some aquaculture practices in East Asia have grown up around possibly invasive species, which can endanger the well-being and survival of native species. (Hodge, I. 2003). The ability of producers and regulators to reduce pollution has been hampered by the lack of available land for agriculture and the persistent dominance of small farms. In several circumstances, insufficient farming acreage has interfered with effective environmental management. Because of the limited chances for market reward and the lax implementation of waste management standards, producers in the animal husbandry industry, for instance, have been discouraged from allocating land for waste treatment facilities. Even while livestock and aquaculture enterprises collectively produce enormous amounts of pollution, regulators may not have imposed stricter waste management standards because of the modest scale (and low capacity and resources) of the majority of such operations. To make ends meet, farmers with limited land are increasingly turning to non-farm sources of income, which can leave them with little time for shrewd farm management techniques (Kraemer, R. A., Choudhury, K., & Kampa, E. 2001)

8.2.4 Compromised food safety

Agriculture is a factor in the concerns of many East Asian consumers regarding the quality and safety of the foods and beverages they eat. Certain agricultural contaminants that enter food by direct deposition and adsorption (adhesion) or bioaccumulation in the food chain via the tissues of animal flesh are a matter of concern. In this regard, one of the main environmental issues China is dealing with is the contamination of its soils with pesticides, heavy metals, and other chemical compounds.

Concerns regarding food safety have also been raised by studies of local food markets in Vietnam and the Philippines, albeit these studies often have limited temporal and geographic scopes. For instance, a study of goods from two communes in the Red River Basin in northern Vietnam discovered alarming amounts of pesticides in samples of vegetables and fish. China has seen a sharp increase in consumer sensitivity to dangers to food safety, and other nations in the area are following suit. However, there is no evidence about the type and quantity of food contamination resulting from agricultural methods, enabling the media to focus on food safety as "scares." While this is going on, a substantial body of research has generally linked both acute and chronic pesticide exposure whether through the consumption of food, water, or other means, to increased rates of chronic diseases like various cancers and diabetes; neurodegenerative diseases like Parkinson's, Alzheimer's, and amyotrophic lateral sclerosis (ALS); birth defects; and reproductive disorders. Some of these endocrine-disrupting and cancer-causing effects may also be brought on by exposure to chemicals in plastics, a problem that is particularly important in China.

Farmers used to raise as many animals as their land would allow in the past. The waste from the crops was added to the natural diets of the cattle, sheep, pigs, poultry, and other animals as well. As a result, the animals helped to maintain the health of the farm as well. Currently, cattle are raised in confined spaces, fed artificial diets, and routinely transported to slaughterhouses. As a result, they contribute to agricultural contamination by emitting pollutants.

8.2.5 The emergence of Drug-Resistant Microbes

Drug-resistant microbes may have already started to emerge as a result of the routine use of antibiotics and antimicrobial agents in the pig, chicken, and aquaculture operations in many countries, including by the majority of small producers. The spread of these microbes through animal flesh, as well as through drinking water and the air, poses a threat to the ability of medications to effectively prevent or treat disease. In 2013, it was estimated that animals in China received 52% of all antibiotic prescriptions.

Significant worries about the emergence of drug-resistant microbes are being raised by the widespread, primarily prophylactic (nontherapeutic) use of antimicrobial agents in overcrowded livestock and aquaculture farms, among other places. Antimicrobial use and environmental exposure over time may place selective pressure on microbial populations in animal guts or aquatic environments, causing the microbes to evolve treatment resistance. At this time, studies conducted in the United States show that there are bacteria that are resistant to five or more medications used in the meat industry (FDA 2012).

According to studies, a significant portion of pathogens sampled in Vietnam's livestock, children living in rural areas, and aquaculture products and waters have become resistant to multiple, sometimes three, or more antibiotics used by the industry. Both humans and animals who are exposed to drug-resistant pathogens should be concerned about this. In addition to airborne dispersal near farms and drinking "tainted" water, human exposure to drug-resistant bacteria can also happen when people consume animal products. Additionally, antibiotics and disinfectants may still be present in trace amounts after wastewater has undergone treatment. Aquatic life native to the area is also susceptible to transmission by translocated aquaculture species. Drug resistance's effects on East Asian populations have not been sufficiently researched. At least 2 million Americans contract an antibiotic-resistant bacterial infection each year, and 23,000 of them pass away directly from their infections; many more pass away from illnesses made worse by antibiotic-resistant infections. The United Nations designated antibiotic resistance as a "crisis" in 2016, using a designation that had previously been reserved for HIV and Ebola due to concerns about its global effects. A 2017 World Bank report on the subject emphasizes the need to phase out the nontherapeutic use of antibiotics in agriculture, citing predictions that antimicrobial resistance could cost the global economy between 1.1 percent and 3.8 percent of GDP by 2050 (Kraemer, R. A., Choudhury, K., &Kampa, E. 2001)

8.2.6 Air quality impairment

Studies from the East Asia region suggest that rural populations and farm workers are most exposed to the health effects of agricultural pollution, including air pollution. For instance, although urban populations are also impacted, sometimes across international borders, agricultural burning exposes farmers and rural communities to the most smoke. Because they lack or do not use protective equipment, or because they use techniques like broadcast spraying that expose the entire rural area, farmers and their families in Vietnam and the Philippines are occasionally directly exposed to the pesticides they apply.

The most important health risks associated with agricultural pollution, however, may go unnoticed because of their gradual onset, enduring nature, and difficulty in observation. When they affect populations of agricultural workers that are poorly organized or have no political representation, even acute health effects are likely underreported. Youth, the elderly, and other populations with lower immunity are actually among the most vulnerable, according to the larger literature on the health effects of pollution.

The health effects of agricultural pollutants may be even more severe for wildlife and ecosystems on the front lines of exposure than for people. China's intensively farmed regions emit reactive nitrogen at high rates, fertilizing non-agricultural ecosystems at a rate that has observable effects on non-agricultural ecosystems. Numerous ecosystem functions, including biodiversity, are known to suffer greatly as a result of this phenomenon.

8.2.7 Enteric fermentation

Cow burps and gas are not fancy phenomena. In the digestive system of ruminant animals like cows, sheep, and goats is an enzyme called enteric fermentation. Gut microbes break down and ferment fibrous foods, such as grass, to create methane, which has a 28–34-fold greater global warming potential than carbon. The largest source of greenhouse gas emissions in agricultural production, enteric fermentation contributes almost 179 million metric tons of emissions that are equivalent to carbon dioxide (Hodge, I. 2003).

The main sources of methane are rice cultivation, livestock rearing, and agricultural burning. As of 2014, rice cultivation and livestock rearing ranked as the region's second and third-largest sources of agricultural greenhouse gases, respectively (based on FAOSTAT data). Agriculture's contributions to ground-level ozone and fine particulate matter, particularly black carbon, and their formation are also potential contributors to near-term warming. Agricultural burning, manure management, and the use of fertilizer and pesticides are the main sources of these emissions. In East Asia, China is by far the biggest producer of agricultural greenhouse gases. Including indirect emissions from intensively grown feed crops, the livestock industry was the largest source of agricultural greenhouse gas emissions in the nation as of 2014 (Based on FAOSTAT data). As of 2010, it was estimated that the use and production of synthetic fertilizers were responsible for a full 7% of all national greenhouse gas emissions. Although estimated to account for 6% of all emissions in 2014, China's agricultural sectors' direct emissions are getting smaller and smaller (based on WRI CAIT data). Particularly, industrial process emissions overtook agricultural emissions after 2005, dropping agricultural emissions to third place (based on WRI CAIT data). However, China's largest source of nitrous oxide and methane emissions comes from agriculture.

Methane levels in the atmosphere are rising, which has prompted researchers to look into its origins. 250 to 500 L of methane can be produced daily by ruminant animals. Based on this level of output, it is estimated that cattle may contribute 2% or less to global warming over the next 50 to 100 years. The amount of feed consumed, the kind of carbohydrates in the diet, the processing of the feed, the addition of lipids or ionophores to the diet, and changes in the ruminal microbiota are only a few of the variables that affect methane emissions from cattle. Cattle methane emissions can be decreased by manipulating these variables.

There are numerous methods for measuring methane emissions from particular animals or groups of animals. Although exact, enclosure techniques call for trained animals and may restrict animal movement. Effective use of isotopic and non-isotopic tracer methods is also possible. Methane output has been estimated using prediction algorithms based on fermentation balance or feed parameters. Although these equations are helpful, their ability to estimate methane output properly is constrained by the assumptions and requirements that must be met for each equation. Mass balance, micrometeorological, or tracer techniques can be used to measure the methane produced by groups of animals. These methods may measure the methane emissions from animal cages, whether they are indoors or outdoors. By utilizing these methods and being aware of the variables that affect methane generation, mitigation measures to lower methane losses by cattle can be developed. By putting these tactics into practice, animal productivity should increase and cattle's contributions to the methane budget in the atmosphere should decrease.

The largest agricultural contributor to greenhouse emissions in the world is cattle. A single cow will belch around 220 pounds of methane annually. According to Mitloehner, a professor and expert on air quality in the Department of Animal Science, methane from cattle has a shorter lifespan than carbon dioxide but is 28 times more potent in warming the environment.

Advocates are asking people to eat less beef because of the growing effects of climate change. They claim it is an unsustainable diet in a world where there will be about 10 billion people by the year 2050. According to them, only 4% of the greenhouse gases produced in the United States are accounted for by cows and other ruminants, and only 2% are directly attributed to beef cattle.

The effectiveness of livestock production in the United States has improved thanks to better breeding, genetics, and nutrition. 140 million head of cattle were required to meet demand in the 1970s. Now, 90 million heads are all that is needed. Additionally, those 90 million cattle are producing more meat at the same time (Kraemer, R. A., Choudhury, K., & Kampa, E. 2001)

8.2.8 Aquaculture water treatment

The majority of modern fish culture in many countries involved a heavy nutrient input in the form of fertilizer or feed, but only a small portion of these nutrients are converted into the target product. The majority of the nutrients are lost to bacterial degradation, which could result in nutrient and methane pollution. For instance, in a laboratory experiment carried out in the Philippines, only one-third or less of the nitrogen and phosphorus ingested in formulated feeds was retained in fish; the remainder was lost through excretion. Since low-quality feeds quickly decompose uneaten in aquaculture waters, more nutrients are lost in systems that use them. Because open systems now predominate in the aquaculture subsector and require the release rather than recycling of excess nutrients, nutrient pollution is a particular concern.

Many nations' aquaculture operations also use a significant amount of potentially harmful chemicals, such as persistent toxic compounds. Pesticides are used to control algae, snails, and other pests, while disinfectants like chlorine and Ozonation are used to reduce pathogens in water. Following the widespread destruction of shrimp farms by the luminous bacteria Vibrio in the Philippines and other nations, many farms started treating inflowing water with chlorine or formalin. To encourage the volatilization of ammonia, water conditioners like lime are added (another pollutant that is a by-product of decomposing feed, feces, and dead organisms).

These processes (pigments, stabilizers, pesticides, and others) pose a particular risk to aquatic biota because they can lead to toxic and occasionally bio-accumulative releases. Some cause harm to humans by contaminating water or eating animal products (Hodge, I. 2003).

Case study

With an average annual growth of 3.2%, aquaculture is the world's largest food production sector. According to reports, during the year 2014, wild fish production decreased more than farmed fish consumption increased. Due to the depletion of the wild stock, aquaculture has recently taken on a significant role in the seafood industry and is now acknowledged as the primary source of seafood for human consumption. The proper implementation of aquaculture systems, particularly in developing nations, ensures food security. The percentage of fish produced for human consumption that comes from aquaculture is anticipated to rise from 52% (the average for the period of 2016–2018) to 58% in 2028. With a contribution to the GDP of 1.07%, India ranks second in the world for the volume of fish produced through aquaculture. According to a recent estimate by the National Fisheries Development Board, the fishery industry in India has amassed a capital of Rs 334.41 billion in the wake of the Economic Revolution through Blue Revolution in the nation. The main problems the world faces today, especially the developing nations, are a lack of water and the depletion of natural water resources. The efficient supply of food depends on the availability of essential resources like energy and water, as well as environmental quality, both of which must be preserved. The production of aquaculture is expected to decline in the future due to several issues, and one of the main obstacles is the availability and accessibility of high-quality water.

This calls for the development of environmentally friendly technologies for the treatment of aquaculture wastewater. Numerous studies have been done to help achieve the sustainability of seafood; the majority of them concentrated on the production area, like nutrient use efficiency in seafood production. Global efforts are being made to ensure that seafood is easily accessible, so an improved seafood industry is crucial to promoting the use of this renewable resource and minimizing the associated environmental effects. The framework for analyzing the relationship between seafood, energy, and water was created by Liu and other researchers to provide a thorough analysis of the activities and resource use in the seafood supply chain. This study assessed the use of energy and water throughout the entire seafood supply chain, including handling, storage, processing, packaging, distribution, marketing, and consumption, as well as the effects these activities had on the environment. The study points out that the quantity of total energy and water consumed in the seafood supply chain with less environmental impact is essential to maintain sustainability.

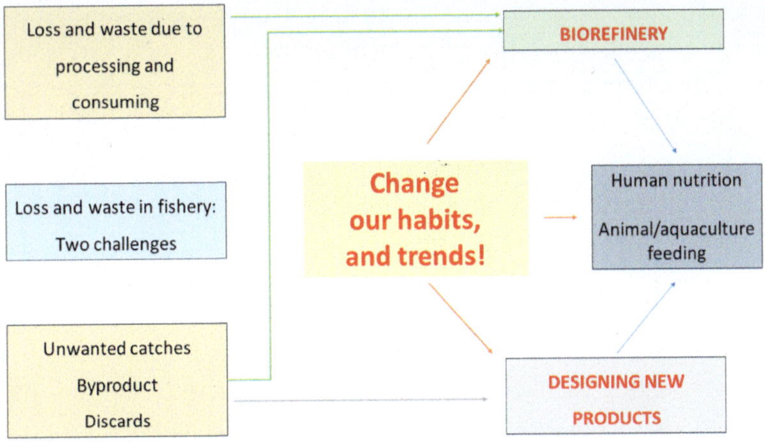

Figure 8.3. Smart aquaculture.

Source: Racioppo A, Speranza B, Campaniello D, Sinigaglia M, Corbo MR, Bevilacqua A. Fish Loss/Waste and Low-Value Fish Challenges: State of Art, Advances, and Perspectives. Foods. 2021; 10(11):2725. https://doi.org/10.3390/foods10112725

Aquaculture wastewater (AWW) pollution of waterways and unprecedented overfishing of wild fish populations for use as ingredients in fish feed raise environmental concerns. The organic components in aquaculture effluents like uneaten feed and feces cause environmental deterioration of both the receiving water bodies and sediments. Dissolved and particulate organic matter, total dissolved solids, and nutrients such as phosphorus and nitrogen are the main component in aquaculture wastewater that can create environmental issues as well as negative impacts on fish growth. The increased metabolic activity of aerobic bacteria is the main result of the significant amounts of decomposable organic waste in aquaculture wastewater. Oxygen scarcity arising from the bacterial action leads to the destruction of the most vulnerable aerobic life forms in the receiving water bodies like rivers and lakes. Aerobic nitrification and denitrification processes are simultaneously inhibited by the anaerobic sulfate reduction reaction that predominates in anaerobic systems. Because sediment macrofauna is necessary for bio-irrigation in natural water streams, which further reduces aeration levels in receiving water bodies, oxygen deficiency in both processes will result in their death. The ecosystem is severely harmed by the organic enrichment in river sediments caused by insufficient microbial population. Such an ecosystem is dominated by sulfate reducers and methanogens, which also disrupt the river basin's connection to the subsequent food chain. To improve the sustainability of aquaculture units, all these environmental effects of discharging untreated aquaculture wastewater to receiving water bodies should be minimized.

The nutrient enrichment of aquaculture wastewater through the feedstock can also result in the eutrophication of lakes and discharge channels, which also requires environmentally friendly treatment methods. In aquaculture systems, conventional wastewater treatment is also carried out using physical, chemical, and biological techniques. For the oxidation of organic matter, nitrification, or denitrification, biological processes like submerged bio-filters, trickling filters, rotating biological contactors, and fluidized bed reactors are used. Traditional wastewater treatment techniques are expensive in terms of initial outlay, energy use, and upkeep. The creation of sustainable technologies is crucial in this scenario for the treatment of aquaculture wastewater. The latest technological options for treating wastewater from sustainable aquaculture have been thoroughly examined in the review, which is necessary to maintain a sustainable connection between the water, food, and energy chains. To ensure the security of the world's food supply, it is essential to make this original comparison of the various aquaculture wastewater treatment technologies based on sustainability (Kraemer, R. A., Choudhury, K., & Kampa, E. 2001)

8.2.9 Cultural belief challenge

Many rural societies view new techniques with hostility and occasionally suspicion. Respect for elders frequently results in the belief that traditional methods are superior. Farmers fear criticism for acting differently from other farmers in addition to the unknown and the untried. The intentions of change-seeking individuals like extension agents are frequently misunderstood in such circumstances. The extension agent may be perceived by the villagers as making changes for his gain. Such viewpoints explain the actions of farmers who appear to agree that a novel approach is worthwhile but are unwilling to implement it (Hodge, I. 2003).

The community has expressed two main points of view: first, recombinant DNA technology is viewed as a powerful tool for improving crop productivity (first-generation GMOs) and food quality (second-generation GMOs), or as "drug factories," for the production of vaccines and/or therapeutic medicines (third generation GMOs). Supporters of GMOs cite data showing that GMOs are necessary for promoting sustainable agriculture because they may be able to lessen the environmental impact of agriculture by reducing the use of pesticides, conserving fossil fuels, lowering CO2 emissions, and preserving soil and moisture. Supporters view GM crops as essential for addressing the serious global food and nutrition security issues in developing nations. Although GM crops have not been hailed as the "absolute solution," it has been asserted that they could unquestionably make a significant contribution to a variety of measurements and incentives to this continuously expanding issue.

Opponents, on the other hand, countered that the side effects in terms of potential harmful effects on the environment and human health are still largely unknown and will likely remain so for decades and encouraged waiting for the final results of additional research and use. The ability of a GMO to escape confinement and potentially introduce engineered genes into wild populations, the persistence of the gene after a GMO has been harvested, the susceptibility of non-target organisms to the gene product, the instability of new genes, the reduction of the range of other plants leading to a significant loss of biodiversity, and an increase in the use of chemicals in agriculture are just a few of the environmental concerns that have been raised. Regarding human health, the main worries have been the potential for allergens to be transferred into the new foods, and the possibility of GM foods transferring their genes to human cells or to bacteria in the digestive tract, which can be concerning. Particularly transferred genetic material was found to harm human health. The mixing of GM crops with those produced from conventional seeds, as well as the introduction of genes from GM plants into conventional crops or closely related wild species, could have an indirect impact on food security and safety.

The same discussion also takes place on a societal level. GM crops will unavoidably fail in the market without "approval" by society as a whole. Therefore, the coming years will be crucial for the application of GMOs in agriculture and food production that is both commercially and economically viable. There are significant differences in consumer attitudes toward genetically modified foods, particularly between North America and Europe. As a result of worries about GMOs, there is information asymmetry, incomplete information, and uncertainty. The biggest worries are concentrated in Western Europe, where consumers' resistance to GMO production and use has grown significantly as a result of widespread ignorance about how our food production system developed, vocal opposition from activist groups, and a steady stream of unfavorable media coverage. Politicians quickly supported this viewpoint. In the 1990s, several food "scandals" that were unrelated to GM foods but cast doubt on the reliability of risk assessments regarding consumer health and environmental risks, with a focus on long-term effects, significantly reduced consumer confidence in the safety of food supplies in Europe.

Since consumers did not perceive any immediate benefits from eating foods made from GM crops, the risk side of the risk-benefit equation received the majority of public attention. Every society's members think that their way of life is the best. "These new farming techniques may be beneficial to some people, but they are harmful to us." This mindset makes people reluctant to try new things. Extension agents may encounter responses such as "How can it be better than our way?" and "We know what is best for us." to suggestions for change.

In conclusion, there are many challenges in the fight against agricultural pollution. Governmental and non-governmental organizations can only do as much as they can to curb pollution but they have to keep up with these problems and also come up with varied decisions on the best steps to accelerate the fight against agricultural pollution.

Chapter

9

THE FUTURE OF AGRICULTURAL CONTAMINANT MANAGEMENT

Contents

9.1 Introduction.. 170

9.2 Strategies and Implementation 171

9.1 INTRODUCTION

Environmental pollution is an important concern nowadays. The majority of environmental toxins are absorbed by soil because it functions as a "universal sink." It contains several contaminants. To maintain soil fertility and boost agricultural output, it is vital to manage soil contamination.

Pollution is the alteration of the physical, chemical, or biological features of the air, water, or soil. This transformation affects human health, the health of other valued living things such as plants and animals, the growth of industry, the quality of life for people, and cultural artifacts. A pollutant is a material that is detrimental to the environment, human health, or quality of life. The majority of pollutants are by-products or leftovers from the manufacture of anything significant, or they are released accidentally or through sewage, rubbish, or other sources. As a result, our vital natural resources, such as the air, water, and soil, are being contaminated.

The soil is the cornerstone of agriculture. It is required for all edible and animal feed crops. This valuable natural resource is being depleted in part as a result of escalating erosion. In addition, "huge volumes of human waste, sludge, and other products" from new waste treatment plants, as well as contaminated water, are creating or contributing to soil contamination. To preserve soil fertility and productivity, which will promote the health of all living things, stringent management systems must be implemented (M. Naeem, Abid Ali Ansari, & Sarvajeet Singh Gill, 2020).

Evaluating the ecological risk of contaminated soil, pesticide usage, sewage sludge amendment, and other human activities that expose the terrestrial environment to hazardous compounds is a difficult and intricate process. Since the middle of the 1980s, terrestrial ecological risk assessment is a relatively young field of study that has evolved rapidly. Soil is normally private property and is sold as real estate, whereas the majority of aquatic ecosystems are often public. Scientists, stakeholders, authorities, engineers, managers, attorneys, nongovernmental organizations (NGOs), and regulators typically have distinct professional and commercial interests. Even if we disregard these elements, the way we now manage risk and the influence of human chemicals on the terrestrial environment presents several issues.

The need for food increases as the world population grows. On the other hand, existing agricultural methods hurt the ecosystem in four main ways. First, fertilizer runoff, which occurs when precipitation deposits an excessive amount of nutrients in lakes and rivers, leads to eutrophication (the overgrowth of algae that deplete bodies of water of oxygen). Second, pesticides threaten ecosystems and water quality. Animal feces that enter the drainage system spread diseases that affect groundwater and fishing spots. Lastly, sedimentation, which occurs when particles settle or suspend in a fluid as a result of erosion, harms fish because silt clogs their gills and kills aquatic plants because sunlight cannot penetrate the murky water.

To encourage farmers to take action to mitigate the effects of runoff, pesticide poisoning, and erosion, governments should launch an educational campaign to raise awareness of the dangers of excessive fertilizer and pesticide use and to provide farmers with methods for reducing nutrient-laden runoff, such as split fertilizer application. The government might also support initiatives that decrease the discharge of pesticides and fertilizers, such as the construction of artificial islands (M. Naeem, Abid Ali Ansari, & Sarvajeet Singh Gill, 2020).

9.2 STRATEGIES AND IMPLEMENTATION

If a national environmental agency does not already exist, this is the initial step. If there are no laws limiting the appropriate use of pollutants in the country, legislation must be enacted to provide the environmental agency the authority to do so and to subsidize activities that decrease the discharge of pesticides and fertilizers. In addition, they will be responsible for educating farmers and community elders on sustainable and efficient agricultural techniques.

The Orissa Social Forestry project in Sweden from 1983 is an example of government and community collaboration. Instead of just asking farmers to grow crops between trees to conserve the forest, government authorities, village leaders, and farmers collaborated to design a forest conservation plan, with some of the ideas coming from the peasants. This might also be used in agriculture, where farmers may adopt new practices voluntarily instead of being ordered to do so, and there could be collective debates about the hazards linked with sedimentation or pesticides.

Once established, the environmental agency will be able to enact relevant rules. A year may be required to start an agency. The educational campaign should be pushed and instructors should be prepared a year before the requirements are implemented. In other words, organizations can begin teaching farmers once they have eligible individuals. This applies anywhere there are farms. Depending on a country's budget, the price fluctuates significantly. The government may decide how much to spend on subsidizing ethical conduct, and the salaries of each agency's employees would be comparable to those of other government workers (M. Naeem, Abid Ali Ansari, & Sarvajeet Singh Gill, 2020).

Reporting norms and standards for contaminated land

The objective is to determine if the property has been contaminated and, if so if the contamination poses a significant risk of harm. The strategy that is employed most commonly consists of three separate reporting periods. Phase 1 begins with a desktop and site walkover assessment with the limited investigation and testing, followed by a Phase 2 detailed invasive investigation and testing for site characterization, and finally a comprehensive Phase 3 report with an evaluation of remediation objectives and a proposed remediation plan, with control and monitoring measures for the activities.

The reporting system necessitates strict adherence to norms and standards of practice so that assessments of the polluted condition of sites can be made most advantageously (considering ecological, social, and economic factors, as well as timeframes). Urgent priority work may necessitate accelerating the stepwise procedure to a contemporaneous single report in some circumstances.

Requirements for preliminary site assessment

When considering a possible location, the following considerations must be made.

1. Description of the site, including its location and size.
2. The kind and magnitude of contamination, the presence of pollutants of concern, and any previous actions that may have impacted the environment. List all current and previous site operations involving the storage, manufacturing, use, handling, or disposal of potentially harmful hazardous materials.
3. Describe the site's present state, its contents, and the results of any past assessment reports.
4. Terrain and geology, drainage, topography, and vegetation of the region.
5. Quality of groundwater and projected water table depth.
6. Near-surface water.
7. Accessibility to sources of potable water.

8. Annual precipitation and the risk of flooding.
9. The utilization of land and water and proximity to the site.

Any future restrictions imposed by the ministry

It is essential to identify any knowledge gaps and assess whether further site characterization data is required, given that not all data may be accessible or that data may vary in terms of uncertainty. Consequently, phase 2 research may commence. The Phase 1 report must provide specific comments regarding the pollution risk at the location. If a thorough site history demonstrates that site operations do not constitute a contamination risk, no further study is necessary and the site should be recommended for reuse. In most instances, a preliminary inquiry will be required to evaluate the level of certainty required to authorize the transfer or redevelopment of real property. To get a waiver on the contaminated status of a site at the Phase 1 level of reporting, it would be essential to perform a restricted examination of certain subsurface operations, such as underground storage tanks (Mateo-Sagasta, 2017).

Approval requirements

The permit holder is obligated to implement a soil management program if the concentration of pollutants in the soil exceeds the appropriate levels specified in the approval. The application must initially address source control to prevent recurring infection releases. After the contamination sources have been eliminated, additional evaluation and delineation of the contaminated region may be necessary. Once the degree of pollution is determined, the following cleanup objectives must be established for the region. When the appropriate treatment or containment technologies have been selected and cleaning objectives have been agreed upon, the management plan may be finalized.

Developing remediation objectives

There are several methods for developing remediation objectives, ranging from simple recommendations to site-specific risk assessments. The broad guidelines are provided as numerical concentration limits and incorporate a range of site situations. When neither the Environmental Protection Agency nor the organization gives rules for a particular item, four possibilities are possible. First, the cleaning aim may be established by the concentration of the ambient background at the site. Second, the procedure of developing recommendations may be implemented. Thirdly, a remediation goal from another jurisdiction may be allowed provided the applicant can demonstrate that it is consistent with the environmental protection objectives of the permission. Lastly, the proponent may develop a remediation objective using site-specific risk assessment methods. The primary focus of a facility's risk assessment will be on human health problems, but basic environmental considerations must also be considered. A site-specific risk assessment determines the possibility that soil contamination may cause harm, given the site's specific conditions. The core components of risk assessments for human health and ecological systems are the same; however, ecological risk assessments are more difficult than risk assessments for human health systems since a diverse array of receptors may need to be considered. The following techniques or components comprise a site-specific risk assessment.

As part of issue conceptualization, a conceptual model of the possible impacts of contaminants on local receptors is created. The conceptual model illustrates how pollutants are dispersed and accumulated at the location of receptors and how these receptors are activated. In the exposure evaluation, the routes that soil pollutants may be absorbed by the receptor are discussed. This information is paired with receptor properties for forecasting contaminant absorption. The examination of toxicity considers the potential adverse effects of pollutants as well as the concentration at which they appear (Mateo-Sagasta, 2017).

Following exposure assessment, the risk characterization phase compares this dosage to the estimated absorption rate to establish the likelihood of a bad result. Risk assessment methodologies can also be used to quantify the amount of a contaminant in the soil where no detrimental consequences are anticipated. On occasion, remediation at a scale consistent with industrial land-use objectives might be impractical. In these circumstances, the permit holder is obligated to guarantee that no receptors are exposed to the contaminants. To achieve these standards, formal risk assessment methods are necessary. Typically, a specified containment system is necessary, and the system's effectiveness must be certified by routine inspections and monitoring. The permit holder is responsible for the design, development, assessment, and maintenance of the risk management system, as well as any necessary emergency reaction, should the system break down.

Enforcement policy

In most cases, enforcement is the final phase in a regulatory process that tries first to prevent prospective issues from occurring and then to collaboratively resolve those that have already happened. When negotiations between the permit holder and the Department fall short of preventing or resolving a violation of approval or the Act, the Department has several enforcement and abatement mechanisms available, including environmental protection orders and warning letters.

Sanctions include citations, administrative fines, orders for enforcement, legal proceedings, judicial rulings, and the cancellation of permits or certificates.

Numerous factors, including the high cost of remediation interventions, necessitate the use of a comprehensive, tiered, risk-based approach based on worldwide best practices to manage remediation in a standardized manner across the nation. To protect both public health and the environment, this is true regardless of the industry. The Framework is based on a review of remediation policies emerging from developing nations and rich nations, as well as an evaluation of potential methods and techniques that may be used for the development of a remediation framework.

Intensive farming

Intensive farming is a method of agricultural intensification and automation that maximizes productivity per acre by using synthetic inputs such as insecticides and chemical fertilizers. Due to the expansion and industrialization of livestock production, millions of animals, such as cows, pigs, and chickens, are now housed in what are often known as "factory farms." Intensive farming practices have become the greatest threat to the global environment as a result of the loss of ecosystem services and global warming, which has resulted in the emergence of new parasites and the reemergence of parasites previously believed to be "under control" by fostering their growth (Mateo-Sagasta, 2017).

By allowing more people to eat from each acre of land and each animal, intensive farming techniques have helped to maintain a growing human population and may have lessened the need to convert the adjacent area into agriculture. Intensive farming also diminishes biodiversity, promotes genetic erosion of crop and animal species globally, disturbs natural ecosystems, depletes and degrades the soil on which it depends, pollutes runoff and clogs water systems, and increases the danger of flooding. According to the World Wildlife Fund (WWF), "agricultural practices, livestock, and land clearance for agriculture are the greatest threats to biodiversity" due to the harm done to beneficial insects and plants by intensive farming and the degradation and depletion of the soil on which these organisms depend. However, intensive agriculture has offset some of the consequences of climate change by increasing yields on previously cleared but potentially underperforming land. Even while intensive farming has certain disadvantages, the benefits vastly outweigh the disadvantages. The earth no longer plays its usual function of delivering an ample food supply. 40% of the planet's surface is devoted to food production, up from 7% in 1700; by 2050, food consumption is projected to have increased by 70% (It is therefore not only possible, but also extremely desirable to feed the world's population with minimal damage to the environment, animal welfare, and human health) (Food and Agriculture Organization of the United Nations, 2018).

Reduce overgrazing

Overgrazing refers to the excessive eating of plants by animals over a prolonged period without a suitable interval of recovery. Common causes of overgrazing include inadequate grazer rotation, an unsustainable animal population on the land, and grazing during inconvenient times of the flora's productive cycle. Overgrazing refers to the overpopulation of wild or domesticated animals, which is a major contributor to worldwide desertification and climate change and impacts one-third of all rangeland. Most commonly, the phrase refers to domesticated grazing animals that are under human care, such as cattle, sheep, and goats. In addition to causing soil erosion, overgrazing diminishes the land's productivity, biodiversity, and utility. It can also cause soil compaction, a decrease in long-term grazing productivity, the loss of topsoil, and an increase in the frequency and severity of surface runoff and flooding.

Monoculture.

Modern food production is predicated on monoculture or the practice of tenaciously cultivating a single crop type across a large area. The technique lacks diversity since the chosen crop, which is often maize, wheat, soybeans, cotton, or rice, is typically replanted in the same soil region year after year without the introduction of a new crop. Benefits of monoculture include less plant competition, decreased waste from faulty planting and harvesting, control of unproductive organisms, and the ability to standardize output. In contrast, monoculture farming rapidly depletes soil resources, relies heavily on synthetic fertilizers and pesticides, and increases the spread of disease when a sensitive crop is grown uniformly. According to research published in Nature, it is beneficial for disease prevention to plant a variety of crops in the same area. According to research done in China, farming many types of rice on the same field enhanced yields by 89%, in part because the incidence of disease decreased by 94%, hence lowering the need for pesticides.

Irrigation.

Irrigation is the practice of artificially applying water to land for a variety of uses, including agricultural production, landscape maintenance, usage in desert locations, and during seasons with little precipitation. According to the FAO, irrigation schemes typically draw water from a river or extract groundwater and distribute it over the irrigated area, which can degrade the land through salinization, alkalinization, waterlogging, and soil acidification; reducing socio-economic conditions by increasing the incidence of water-related disease; create inequity, and weaken community infrastructure; create poor water quality by reducing irrigation water quality, and create wetlands. In addition, irrigation may contribute to ecological degradation by lowering biodiversity, damaging downstream ecosystems as a result of decreased water quantity and quality, and depleting groundwater by decreasing base flow and wetlands and causing dry drinking and irrigation wells. FAO states that "The advantage of irrigation is that it allows a nation to enhance food and forage production in the most favorable areas while lowering the strain on marginal regions that are now utilized for grazing or rain-fed crops. These types of land are already eroding (known, in semi-arid areas, as desertification). People that have relied on rain-fed agriculture for millennia may find it challenging to adapt to irrigated agriculture on a societal level (Food and Agriculture Organization of the United Nations, 2018). However, severe land deterioration will necessitate modification. If necessary, irrigation development is a viable alternative to starvation or mass migration. Irrigation-related development may have both beneficial and bad environmental consequences. For irrigation to be sustainable, adverse impacts must be minimized."

Slash and burn.

Slash and burn for farming, forests and other natural places must be destroyed, affecting 200-500 million people globally. It is usually associated with rainforests and is used by indigenous for food production (i.e. farming to survive). Following slash-and-burn agriculture, soil loss and erosion can be so severe that regeneration can take decades, if not millennia. Slash-and-burn agriculture may be viable in enormous woods where fields have time to recover before being cut down, burnt, and cultivated, as it has been for nearly 12,000 years. Communities now have access to food and money as a result of growing in formerly inaccessible areas made possible by contemporary technologies. However, because populations are too dense for this technique to work, farmers would have to quit their plots and start again in less populous places, as the soil would deteriorate fast in the absence of trees (Food and Agriculture Organization of the United Nations, 2018).

Figure 9.1. Slash-and-burn agriculture.

Source: Rösch M, Biester H, Bogenrieder A, Eckmeier E, Ehrmann O, Gerlach R, Hall M, Hartkopf-Fröder C, Herrmann L, Kury B, Lechterbeck J, Schier W, Schulz E. Late Neolithic Agriculture in Temperate Europe—A Long-Term Experimental Approach. Land. 2017; 6(1):11. https://doi.org/10.3390/land6010011

Slash-and-burn agriculture in the tropics accounts for almost a quarter of global greenhouse gas emissions. Loss of forest cover, which can be temporary or permanent if fields aren't given enough time for vegetation to grow back after a slash-and-burn technique, soil erosion, which always occurs but is especially problematic when the technique is used on adjacent fields, resulting in water shortages and nutrient loss, and desertification, which occurs when the flora is wiped out due to a lack of water, are all environmental problems that can be attributed to people's actions.

Reduce deforestation and begin reforestation

Deforestation and soil erosion are tightly connected. In the absence of trees, which would normally absorb and sustain these liquids and the accompanying pollutants, acid rain and floods can devastate healthy soil. Soil erosion can occur when there are no trees or few plants to prevent the top layer of soil from being carried away by natural forces such as water and air, which contribute to soil pollution. By reforestation efforts and the planting of new plants in locations prone to erosion, soil pollution may be reduced even more.

Inorganic toxic compounds

Inorganic wastes considerably impede industrial waste disposal. They include metals with a high toxicity potential. Significant quantities of arsenic fluorides and sulfur dioxide are also created by industrial processes (SO2). Fluorides are abundant in the environment due to the industries that create superphosphate, phosphoric acid, aluminum, steel, and ceramics. Soils may become extremely acidic due to sulfur dioxide emissions from industrial and thermal operations. These metals are harmful to plants and leaves (Winder et al, 2002).

Copper, mercury, cadmium, lead, nickel, and arsenic can accumulate in the soil if they are introduced by sewage, industrial waste, or mine washings. Additionally, the soil is poisoned with fungicides containing copper and mercury. Soil particles absorb lead, which is contained in automotive exhaust and is damaging to plants. Toxicity can be reduced by increasing soil organic matter, supplementing soils with lime, and maintaining soil alkalinity.

Avoid soil erosion by planting all over the year

After harvest, cover crops can be used to prevent bare ground, which can cause soil erosion and, as a result, the disappearance of waterways.

Plant trees or grasses along the edges of fields

Planting grasses and trees along the edges of a field is important when the field lies on the borders of water bodies.

Through these planted buffers, nutrient losses can be avoided by filtering out nutrients before reaching the groundwater.

Adjust the intensity of the tillage of a field

Farmers should consider the reduction in the tillage of their fields to reduce runoff, soil compaction, and erosion. Thus, a more sustainable path toward farming and a reduction in agricultural pollution could be accomplished.

Improve manure management

Animal waste contaminates agriculture to a significant degree. Establishing and enhancing methods to combat these contaminants is therefore vital. There are a variety of manure treatment options available to reduce the environmental damage caused by manure. To ameliorate the situation, these strategies must be further developed. To prevent agricultural pollution, farmers must adopt several critical measures, including waste management. Manure may be a vital source of fertility and organic matter, providing essential nutrients such as nitrogen and phosphorus, if applied to crops in the proper amounts and at the proper periods. Several regional governments have begun implementing manure management legislation and incentive programs to reduce nitrogen loads on farms and preserve vital watersheds. Some areas, such as the Chesapeake Bay, have also imposed agricultural nutrient load limitations.

In addition, some farmers have begun utilizing "digesters," which are innovative devices that convert methane from manure into energy, therefore reducing the environmental impact of farms.

Change our consumption behavior

Part of this problem lies in the fact that most of us consume an excessive amount of meat. To meet this demand, farmers have to use large amounts of fertilizer to grow enough animal feed to be able to meet the meat demand. Changing our consumption behavior to less meat consumption can mitigate the problem of agricultural pollution to a certain degree.

Prevent excess nutrients from reaching the water

Farmers can build fences around water bodies to prevent access for animals and thus excessive amounts of nutrients enter the water. Although this is just a small measure to reduce agricultural pollution, it still is a small step to reach our goal of less pollution.

Raise the awareness of farmers on the topic of agricultural pollution

Farmers have to be convinced and educated of the negative effects they may (often unknowingly) cause to the environmental system. They should be taught that the excessive use of fertilizer and pesticides has a huge adverse impact on the whole ecosystem. Thus, by increasing the farmers' knowledge, agricultural pollution can be mitigated to a certain degree.

Education

It is possible to solve any difficulty with enough education. The public has to be made aware of the drawbacks associated with agricultural pollution. This can only be accomplished with proper education beginning at a young age. We can make a big difference in getting people to change their environmental practices if we start early and get kids thinking about the effects of agricultural pollution on the ecosystem. There is a higher chance that these children will be able to convince their parents of the negative consequences their activities have on them.

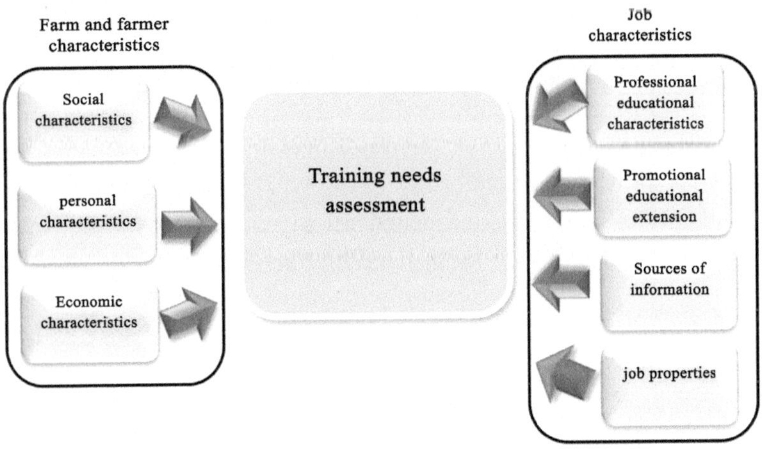

Figure 9.2. Farmers' learning process, motivations, and role of education.

Source: Goli I, Azadi H, Miceikienė A, Tanaskovik V, Stamenkovska IJ, Kurban A, Viira A-H. Training Needs Assessment: The Case of Female Rice Farmers in Northern Iran. Agriculture. 2022; 12(3):390. https://doi.org/10.3390/agriculture12030390

Children's attitudes may have an impact on their purchasing behaviors.

In addition, if youngsters are taught to think critically about the impact of their purchasing decisions on agricultural sustainability, they may grow up to be responsible people. To get people to change their ways and lessen agricultural pollution, education is essential.

Convince others

For the problem of agricultural contamination to be resolved, our participation is necessary. However, our routine actions fall short of making a significant impact. We must also persuade our loved ones and close companions of the environmental harm caused by their consumption. We must demonstrate how they consistently contribute to agricultural pollution. If they do, individuals are more likely to modify their behavior.

To boost the efficacy of this process, we must educate individuals not only on how their behaviors harm the environment but also on potential remedies. In addition to our criticisms, we must provide usable answers that we can use in our everyday lives. For instance, this would involve reducing your weekly meat consumption to two or three servings.

Nutrients

Fertilizers, such as nitrogen and phosphorus, pose several dangers when they seep into groundwater, though the risks vary depending on the nutrient and soil type. Phosphorus nearly never makes its way into groundwater due to its low solubility in water, much less so in sandy, clay-free soil sites. Nitrogen, unlike phosphorus, is water soluble and quickly changes to nitrate, which can contaminate groundwater if not utilized by plants. Nitrate concentrations in groundwater can be reduced by supplying enough nitrogen through methods such as chemical fertilizers and animal dung. Nutrient uptake by plants and crops is dependent on their delivery in the appropriate amount, at the right time, and in the right place. The amount of manure to apply is affected by the nitrogen content of the manure, the amount of nitrogen already presents in the soil, and the crop's nitrogen requirements. A lab test of your soil and manure can tell you how nutritious they are. Contact your county or state's Cooperative Extension Office for a list of soil and manure testing facilities (Winder et al, 2002).

Nitrate can be removed from the ground in two methods. Plants use it as a fertilizer at first. Reduce nitrate pollution by just applying as much fertilizer or manure as plants can use throughout the growing season. Before applying any manure or fertilizer, test the soil once more. Second, because certain bacteria are capable of removing nitrate from damp soils, the threat of nitrate pollution of groundwater beneath these soils is minimized. To reduce contamination once nitrate leaches below the root zone in well-drained soils, fertilizer and irrigation must be carefully managed.

Pesticides

The likelihood of pesticide release is greatest in sandy soils with the least organic content. Because there are fewer microorganisms present, pesticides degrade more slowly in these soils, and leaching can occur rapidly due to the huge soil pores. Even though the coarse texture of these soils cannot be altered, the capacity of pesticides to bind to these soils can be enhanced by adding additional organic matter.

Microorganisms (Pathogens)

The digestive tracts of animals include pathogens (microorganisms), which are discharged in animal waste. If they enter surface water, they can spread illness to cattle and humans. When surface water flows through the soil, physical (filter), chemical (adsorption), and biological (predation and natural die-off) mechanisms protect groundwater against pathogen and bacterial contamination.

Organic Matter

Manure also contains organic components. Bacteria require organic matter for sustenance. Oxygen is utilized when microorganisms break down organic matter. In streams and ponds, microorganisms decompose vast quantities of organic waste, using all the dissolved oxygen in the process. Aquatic life, especially fish, can be negatively affected by low oxygen levels.

Soil and water quality

Before contaminants can reach groundwater, the soil may filter them out. Water percolates through the soil and into the groundwater, where pollutants attach to soil particles. On soil particles, naturally occurring microorganisms degrade pollutants into harmless molecules. The organic matter content of the soil is also crucial. Rich in organic matter, soils can chemically and biologically degrade pollutants before they reach groundwater. Even in regions with ideal geology, poor management practices can pollute groundwater, particularly after heavy rainfall when runoff is possible.

What is Runoff?

Runoff happens when the ground gets more water than it can hold from precipitation or irrigation. Animal-inhabited farmlands may be more prone to a runoff if, for example, they are overgrazed or have inadequate grass cover. When discharge containing bacteria (from manure), dissolved nutrients (from commercial fertilizers and manure), or pesticides pass through the soil and reach the water table, groundwater can become polluted. The rate at which water may permeate the soil is referred to as the infiltration rate. Certain regions next to structures have been compacted by animal or machine activity, resulting in exceptionally low infiltration rates. Here, normal storm conditions result in runoff. Roofs and paved surfaces prevent infiltration. These places accelerate and augment the amount of water that drains from neighboring locations.

Using AGRIVI to combat agricultural pollution

AGRIVI, a digital agricultural system, is another crucial instrument for enhancing farm production and sustainability. Using farm management software such as AGRIVI, farmers may track the amount of fertilizer and pesticide they apply, compiling this information into accessible reports for compliance, record keeping, and data analysis. Farmers may also generate prescription maps that modify fertilizer and pesticide application rates depending on real-time information on crop yield, soil fertility, and plant health. Utilizing weather and pest notifications, farmers can schedule their spraying operations and minimize duplication. AGRIVI farms may adopt preventative methods to eliminate pests, diseases, and abiotic concerns before they become a problem (Mateo-Sagasta, 2017).

Buffer zones and habitat creation

Farmers can also plant "buffer zones," which are carefully positioned plant sections that give a variety of environmental advantages, such as:

- Reduced agricultural erosion and runoff
- Improved water filtering in vulnerable areas (such as along a stream or in an important watershed)
- reducing pesticide vaporization through windbreaks • improving habitat for local animals, among other benefits

Plants can be strategically planted in these places according to their distinct root systems, growth patterns, absorption capacities, or ecological responsibilities. For instance, riparian buffers are a type of buffer system that consists of a variety of trees, shrubs, and perennials to protect watersheds and improve stream quality. Government institutions such as the EPA and USDA promote the planting of buffer zones, typically in collaboration with regional environmental organizations.

Cover crops and conservation tillage

Conservation tillage, which refers to any tillage approach that leaves 30 percent of the soil surface covered with residue, is a typical method employed by farmers to prevent soil erosion on their farms. Leaving residue on the soil surface or planting leftover from the previous crop reduces the likelihood of erosion and runoff during heavy rain events because the roots and plants from earlier plantings act as a barrier between the soil and the wind and rain.

Cover crops, which may be farmed in some combinations throughout the year, operate in tandem with conservation tillage to prevent soil erosion and enhance soil quality. During fallow times, oats, vetch, and rye are examples of cold-season cover crops that are grown. Warm season cover crops such as sorghum, Sudan grass and cowpeas are grown during this period to build soil organic matter, remove excess nutrients, and nourish the soil microbiome. The USDA and other government agencies offer a variety of incentive programs to encourage farmers to implement conservation tillage and cover crops due to their numerous advantages.

Cleaning contaminated soil

Although it is not "actually" possible to eliminate toxins from the soil, there are several measures that may be done to reduce the risk. Bringing the pH of the soil as near to neutral as possible will assist in mitigating the harmful effects of pollutants. To rehabilitate polluted soil, a healthy top dress of peat moss, compost, or aged manure must be used. This method will aid in protecting plants. Always wash your fruits and veggies before consuming them. If pollution is an issue, use untreated wood-raised planters. Therefore, you will be able to add your nutrient-rich soil. Taking the required steps to remove polluted soil before planting can result in a healthier garden for you and your family (Mateo-Sagasta, 2017).

Funding

The environmental agency's budget will be utilized to subsidize the educational efforts. Farmers will be required to make specific adjustments, some of which will be at no cost, such as modifying how they till the soil. Some, such as pesticide rotation, might be excessively expensive for farmers, demanding subsidies from the government. If the government is unable to give subsidies, communities may turn to microfinance, the provision of small loans of up to $1,000 by private companies.

Animal wastes

There are numerous ways to prevent animal excrement from contaminating the water. Farmers should keep clean water from accessing regions containing animal and livestock waste by constructing terraces to limit runoff and streams to redirect water flow. Groundwater and surface water can be protected by keeping livestock at a safe distance from wells and streams and by constructing a fence around them to keep animals away. When utilizing animal excrement as fertilizer, the same measures should be taken. A buffer zone can be utilized to absorb runoff from fertilizers and pesticides.

Controlling runoff by the direct transport of water to a holding pond for later use is possible with a runoff management system. A grass filter strip is another method for collecting runoff and preventing nutrients from entering bodies of water.

Instead of storing manure on the sand, it should be placed on a concrete, clay, or plastic pad to allow water to drain off the mound. If manure is delivered by pipelines and pumps, it should be routinely inspected for leaks.

Liquid waste may be held in lagoons, which are clay pits until microbes break down the organic matter.

Before utilizing animal feces as fertilizer, farmers should verify its nutritional content to avoid over fertilizing crops. This rubbish should be used as fertilizer on days with no precipitation, as little wind as possible, no frozen ground, and less than half an inch (12 mm) of precipitation the previous week. If possible, animal feces should be incorporated into the soil.

Sediment

When water rushes across the land, it causes erosion by pulling up the top layer of soil. Reduce salt in the outflow by minimizing erosion. Lower erosion rates can be attributed to less frequent tilling of organic agricultural soil and the use of a winter legume cover crop. In a conventionally farmed zone, water erosion occurred at a rate of 32.4 tons per hectare (2,939 tons per square kilometer), but in an organically farmed area, the rate was only 8,3 tons per hectare (753 tons per square kilometer). Minimal tilling is one of the best methods for preventing soil erosion. After crops have been harvested, the crop waste can be utilized as a rain barrier. Due to the extensive amount of exposed soil in orchards, more specific methods are required.

Research indicates that contour farming, or growing down the slope rather than up and down, can avoid up to fifty percent of soil erosion.

In the spring, when the winter cover crop has been sown and maintained, it is either mowed or incorporated into the main crop. Grass and legumes are two other superior options.

When organic matter is added to soil, its capacity to store water increases.

By tilling the soil gently, crusts that make it difficult for water to penetrate the soil may be prevented.

Crop Rotation

Farmers should use crop rotation rather than planting the same crop year after year in the same region, as is often done. Even though monoculture is less successful than crop rotation and has negative impacts on the environment and human health, it is widely used because it allows for the planning and harvesting of large areas of land.

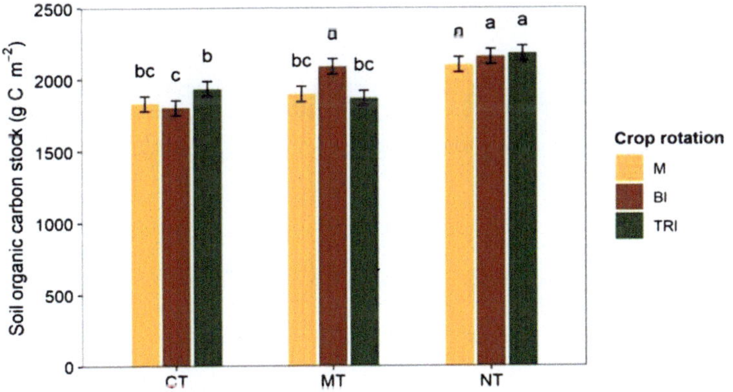

Figure 9.3. Crop rotation management in a farm: Effect of two-way interactions of 'Tillage × Rotation' on the soil organic carbon stock (SOCS, g C m−2). Tillage effect (No till: NT, moderate tillage: MT, and conventional tillage: CT) and Rotation effect (Monocropping: M, biennial: BI, and triennial: TRI). Data are averages ± standard errors. Different lowercase letters indicate significant differences between treatments ($p < 0.05$).

Source: Jaziri S, M'hamed HC, Rezgui M, Labidi S, Souissi A, Rezgui M, Barbouchi M, Annabi M, Bahri H. Long Term Effects of Tillage–Crop Rotation Interaction on Soil Organic Carbon Pools and Microbial Activity on Wheat-Based System in Mediterranean Semi-Arid Region. Agronomy. 2022; 12(4):953. https://doi.org/10.3390/agronomy12040953

The broad planting of a single crop increases the likelihood of insects, weeds, and other pests infesting the plants, making it more difficult to eliminate the problem once it has established itself. Crop rotation, on the other hand, inhibits weed and insect development by requiring them to constantly adapt to new surroundings and nutrients, which is considerably more difficult for them than it is for humans.

Such research has been carried out, for example, on a sugarcane plantation in Australia, where the farmer used a reduced-tillage agricultural approach that included a rotating crop of soybeans. In comparison to the previous farm, the new farm used 8% less fertilizer ($309 per hectare) and spent $47 per hectare on weed management instead of $60, resulting in an extraordinarily cost-effective crop. Soybeans were priced at $539 per acre (Food and Agriculture Organization of the United Nations, 2018).

Crop rotation, the use of pest-resistant plants, and simple preventative steps like washing farm trucks between fields may dramatically reduce weed seed transfer, therefore pesticides should be used only as a last resort. You should not use them if there is a chance of rain or if you have any irrigation plans.

When spraying pesticides, it is advisable to alternate between several varieties. By changing the pesticides that are employed, resistance to such chemicals can be prevented, reducing the need for their usage.

Pesticides can also be applied in a variety of environmentally acceptable methods. By pushing the herbicide into the topsoil, soil integration lowers pesticide runoff. Another method for reducing runoff is early pre-plant treatment, which uses less pesticide than usual and is sprayed directly into the soil before the crops emerge. If this is done, half as much pesticide will be washed away. Weeds that remain after general treatment might be targeted for removal utilizing spot treatments.

Apply 12-23% of the insecticide before planting and 13% after planting to minimize runoff by one-third. Furthermore, if the weeds are managed with the first spray, a second treatment may be avoided (Food and Agriculture Organization of the United Nations, 2018).

Another method for preventing pesticide runoff is to employ an adjuvant, often known as a spray mix component. Adjuvants are chemicals that are used to increase the efficiency of pesticides by increasing their interaction with plant leaves. In one investigation, cucumber and pea leaf samples were used. Administering 0.25 mL/L of an adjuvant to pea plants enhanced retention by more than a factor of four. Cucumber plant leaves are covered in amorphous wax (as opposed to pea plant leaves, which have crystalline wax) and have deep veins that easily channel runoff, so the spray did not work as well on them. Additional study is now being conducted in this area.

In addition to the precautions farmers should take while spraying pesticides, a vegetated buffer strip can be constructed to absorb toxins when runoff exceeds the buffer. Buffer strips can range in length from 9 meters to 15 meters, but their utility is independent of their width. According to USDA authorities, any buffer is preferable to no buffer at all.

Pesticide education is also essential, especially for farmers who may not be able to read the labels. Pesticides must be handled carefully during the storage and disposal of empty containers. Wearing protective equipment, such as a mask, is essential for staying safe while spraying pesticides. Two more common sense precautions are wearing gloves and tucking pants into boots.

Governments should impose strict regulations on the distribution of highly toxic pesticides [6], and these substances should be treated in the same way as other illegal drugs. If this chemical is used, the producers will be fined. Toxic substances were discovered during a pesticide management meeting between the Food and Agriculture Organization and the World Health Organization.

Though all nations should accept some form of government oversight, Mission 2017 recognizes that implementing such laws in less developed countries may be difficult. If there are no regulations in place at the moment, environmental agencies in more developed countries should enact some. These restrictions could be patterned after those used in other countries. Not all pesticides are allowed, and some are even taxed (or, as in the case of Indonesia, ending subsidies, which discouraged pesticide use). If pesticide application certification is implemented, pest control expertise is also likely to be required. Separate regulations in a developed country may require spraying equipment licensing and the use of high-tech, computer-controlled sprayers to provide farmers with a better understanding of the precise amount of pesticide being administered. Pesticide overuse is a widespread problem in many developed countries due to concerns about its efficacy. Rich countries can mitigate some of this volatility by instituting and strictly enforcing a licensing regime. Pesticides could be regulated using adaptive licensing in the same way that pharmaceuticals are. The initial license is granted more quickly in this method, but the pesticide's use is restricted during the trial phase, resulting in a two-stage licensing process.

Fertilizers

Thanks to fertilizers, crops may survive in soils deficient in essential nutrients for plant growth. However, excessive fertilizer can contaminate water with nitrogen, phosphorus, and other elements, causing soil acidification and eutrophication and contaminating groundwater. Efforts should be made to expand understanding of the impacts of over fertilization to prevent difficulties.

Creating artificial floating islands that rely on plants to absorb excess nutrients is one way to mitigate the detrimental impacts of fertilizer. In addition, enhanced agricultural practices such as fertigation, which involves irrigating with water containing dissolved nutrients, and nitrogen sensors, which provide more information about a soil's nutritional needs, can be implemented (Fouché, O., Lasagna, M., & Danert, K, 2019).

Pesticides

Pesticides (herbicides, insecticides, and fungicides) can seep into the groundwater or be carried by runoff into nearby bodies of water, where they can cause cancer, liver and kidney damage, and reproductive issues if consumed. Farmers in both developed and developing nations can take the following methods to decrease pesticide runoff.

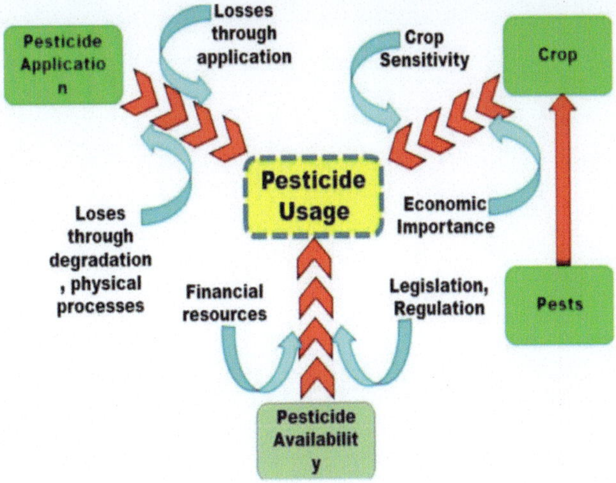

Figure 9.4. Pesticide usage.

source: tudi m, daniel ruan h, wang l, lyu j, sadler r, connell d, chu c, phung dt. agriculture development, pesticide application and its impact on the environment. international journal of environmental research and public health. 2021; 18(3):1112. https://doi.org/10.3390/ijerph18031112

Chapter

10

UNIVERSAL BODIES PUSHING FOR SUSTAINABLE CONTAMINANT-FREE AGRICULTURE

Contents

10.1 Introduction 194
10.2 Terms in Universal Sustainable Agriculture Bodies 196
10.3 United Nations Environment Program (UNEP) 203
10.4 Global Alliance on Health and Pollution (GAHP) 207
10.5 European Environment Agency (EEA) 209

10.1 INTRODUCTION

As human beings and a large percentage of living organisms, we eat to live. Food provides all living organisms with the energy needed to sustain life. For human beings, agriculture is the main activity that has been known to supply almost three-quarters of their day-to-day lives. Agriculture involves the cultivation of edible plants which give their produce either in form of seeds (for instance maize produce maize corns which are used both as food and also as seedlings during planting season); or as leaves, or stems, among other plant parts. It is just to say that different plants grow to provide food to human beings through different parts of the plants.

Agriculture has also been used as an economic activity whereby different crops are planted and later sold in large quantities, earning farmers a source of livelihood. Agriculture does also involve the rearing of different livestock which either serve as food or as a source of energy within the farms. Despite agriculture being an important sector in most parts of the world, it has faced different challenges which affect its success. Such challenges include climate change which has had a negative influence on both planting and harvesting seasons due to changes in both rain and dry periods of the year. The discovery of such challenges has pushed most nations to do thorough research and try to come up with ways that ensure agriculture remains a success at all times (Chhipa, H. 2017).

Sustainable agriculture is one of the main results of the research done in different parts of the world. The main aim of sustainable agriculture is to ensure that all society's food needs are met without causing any long-term effects that might affect the agricultural growth rate shortly. The enthusiasts of sustainable agriculture hope to combine three main objectives in their research which include: the sustenance of a healthy environment, good economic profits from agricultural activities, and the promotion of equity both in social and economic sectors of agriculture. For sustainable agriculture to be a success, every individual involved in agriculture has a main role to play. The food system growers have the responsibility of promoting soil health and ensuring efficient use of water during the planting process and also reducing the levels of pollution in their farms. Food distributors and consumers have the role of ensuring that they only purchase value-based foods that have been grown using environmentally friendly techniques; a move which encourages more farmers to adopt environment-friendly farming practices. As for the researchers, they have the role of exploring different disciplines and testing if they can be combined with agriculture to deliver some amazing results. For instance, research on how technology and engineering could be incorporated into agriculture to automate some agricultural activities.

Sustainable contaminant-free agriculture ensures that all agricultural activities promote ecological balance through avoidance of the use of environment-contaminating materials such as harmful pesticides and herbicides. It is believed that with sustainable contaminant-free agriculture, the issue of food security will be addressed. Agritourism will be introduced in different parts of the world serving as an extra source of income for different farmers. Agroforestry will be a new thing promoting soil health and good climatic conditions in regions where agriculture is being practiced in plenty. Biofuels are believed to be used in replacing harmful farm machine fuels which are not only costly but also end up contaminating the soil in cases of unnoticed spillage. Conservation tillage is to be introduced in most farms that practice sustainable farming hence eliminating the issue of soil erosion in farms. Cover crops could be used for mulching purposes promoting the retention of soil moisture and upon decay serving as a source of manure for the plants. Direct waste management will ensure that the farms are clean promoting a healthy agriculture environment.

With sustainable agriculture, direct marketing is made possible for distributors are expected to purchase food materials directly from the farmer, eliminating the need for a middleman, which in turn leads to great profits for both the farmer and the distributor. Sustainable agriculture requires an intensive labor force, opening a platform for employment to unemployed individuals. Food waste management is a top-list activity in sustainable agriculture promoting recycling processes that help in minimizing the cost of agricultural production. The growth of genetically modified crops is to be regulated to prevent the negative health effects most consumers experience due to the consumption of genetically modified crops. Organic farming ought to become the new norm since it reduces the rates of pollution in farms by a large percentage.

Soil nutrient management ought to be promoted in sustainable agriculture to ensure that the crops being produced are in healthy condition. In most farms, postharvest practices are not well coordinated leading to delayed cultivation. Sustainable agriculture aims at introducing postharvest management practices which will ensure that every farming activity is completed in due time. Technological innovations are being incorporated as the new norm in sustainable agriculture minimizing the excessive labor force required through the automation of most farm activities. With water being the main support system for all agricultural activities; be it planting of crops or rearing of animals, sustainable agriculture aims at ensuring that there is an efficient use of water and promoting quality water management. Zero-emissions freight transport is aimed at promoting a healthy environment in most farms by ensuring that harmful fuel emissions are minimized.

Urban agriculture has become a new norm in most urban centers. However, the main challenge facing urban agriculture is access to quality farming materials such as seeds, organic manure, and a healthy environment, among others. Sustainable agriculture aims to ensure that urban agriculture is made easy by providing all the necessary farming materials to wannabe urban farmers. Urban farming is aimed at ensuring that both the rural and urban centers in different parts of the world are well supplied with food.

Most nations across the world have realized the worth of sustainable agriculture to not only their current generation but also to many generations to come. Several environmentally friendly organizations/ bodies are pushing for full support for sustainable agriculture, not only by the public sector but also by the private sector. The idea of environmental bodies such as the United Nations Environmental Program pushing for sustainable agriculture is a result of the decline in climatic conditions in different parts of the world; something which has adversely affected the agricultural sectors hence negatively reflecting on the livelihoods of a large percentage of the world's population (Chhipa, H. 2017).

The universal bodies that are pushing for sustainable contaminant-free agriculture include UNEP, The United Nations Environmental Program body; GAHP, The Global Alliance on Health and Pollution; and EEA, The European Environment Agency.

10.2 TERMS IN UNIVERSAL SUSTAINABLE AGRICULTURE BODIES

The universal bodies that are pushing for sustainable contaminant-free agriculture agreed on some common terms which are to aid in the promotion of sustainable agriculture across different parts of the words. These terms are discussed below.

Agrarianism: Agrarianism is a term used to refer to either the social or political movement that is aimed at bringing about land reforms and improving the economic status of farmers. Agrarianism, unlike most agricultural terms, sends the message that human beings are the villains in the world of agriculture and the environment in general. The main reason agrarianism was categorized under sustainable agriculture is because it acknowledges that human beings are biological and social beings that rely on healthy environments and communities for their survival. Agrarianism matters to sustainable agriculture since it reminds us humans that we must always be sympathetic to our cultural responsibilities of agriculture which draws them to practice sustainable agriculture.

Agro-ecology: it is used to refer to the combination of an environment-friendly and sensitive approach to agriculture. Agro-ecology is an important term in sustainable agriculture that emphasizes not only food production but also the ecological sustainability of the agricultural field. In terms of sustainable agriculture, Agro-ecology mostly focuses on the study of ecological existences within the crop field such as the predator and prey relationships and the crop and weed competition during their growth life.

Alternative farming: this is one of the most used terms in sustainable agriculture for it introduces different concepts which are aimed at promoting the idea of sustainable contaminant-free agriculture. It does introduce the idea of cultivation of nontraditional crops, rearing of exotic livestock, and use of other exotic farm products. It promotes the idea of quality service delivery in the farms, recreation activities, tourism, food processing, agroforestry, and other enterprise activities which could earn the farmers an extra income and cultivate the spirit of farming among uninterested individuals. Alternative farming roots for organic farming and aquaculture in sustainable agriculture. Direct marketing techniques are a concept of alternative farming, that as stated earlier, help in increasing the profit percentage for both the farmers and the distributors. Generally, alternative farming is all about farm diversification and the benefits expected to be earned from it (Chhipa, H. 2017)

Best management practice (BMP): this is a term used in sustainable contaminant-free agriculture to refer to the establishment of soil conservation practices that are aimed at providing water quality benefits to farms and the surrounding society at large. The mentioned practices include the likes of use of cover crops, the use of green manure crops, and strip cropping that is aimed to help in curbing soil erosion. Soil testing is also done to help in preventing the loss of nutrients in the soil and maintain an ecological balance of the pesticides which reside in the soil. Biodiversity: this is used in sustainable agriculture to refer to the summation of all plants, animals, fungi, and microorganisms in a particular farming area and their interactions with each other. Agrobiodiversity is derived from biodiversity and used in sustainable contaminant-free agriculture to refer to all the biological resources which are associated with agriculture. The resources include genetic resources which aid in the growth of both animals and plants within a farm; edible plants and crops of all varieties; livestock, consisting of both small and large ones and also aquatic farm animals such as fish; soil organisms which help in boosting the soil fertility and generally the soil health; naturally occurring insects, bacteria, and fungi that are known to control diseases of domestic plants and animals; agroecosystem types that are responsible for nutrient cycling, stability and soil productivity; and wild resources which include natural habitats and landscapes that are known to be of great use in agriculture. Generally, agrobiodiversity is all about the exploitation of biological diversity for the growth of sustainable contaminant-free agriculture.

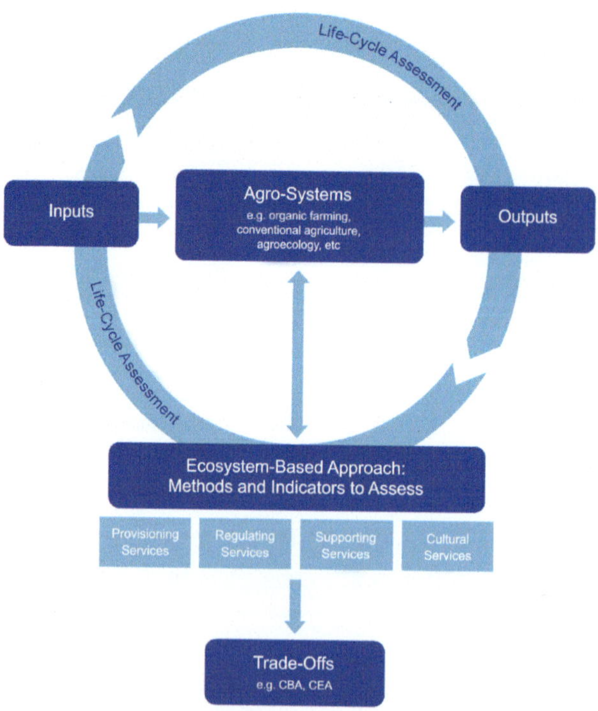

Figure 10.1. An illustration of sustainable agriculture assessment and evaluation.

Source: El Chami D. Towards Sustainable Organic Farming Systems. Sustainability. 2020; 12(23):9832. https://doi.org/10.3390/su12239832

Biodynamic farming: this is a term used by bodies that push for agricultural biodiversity to refer to the concept of describing farm preparation practices in line with parameters that are influenced by forces of living nature. Biological farming, as the name suggests, is sustainable agriculture that involves the growing of crops and rearing of animals with minimal use of chemicals to control crop pests and animal diseases. In most cases, the term ecological farming could be used in place of biological farming; this is because biological farming promotes ecological balance ensuring that everything within a farm environment thrives naturally, by the laws of nature.

Biotechnology: this is an important term in sustainable agriculture. It advances the traditional use of biotechnological concepts in agriculture through the introduction of genetic engineering. Genetic engineering allows the combination of DNA samples from several species to aid in the creation of a completely new organism. With genetic engineering, better breeds of plants and animals are produced giving them the ability to overcome the constant climatic changes which are adversely affecting the survival of both indigenous crops and plants. However, biotechnology is used with caution when it comes to sustainable agriculture, for experts have to ensure that whichever technology they invent, is thoroughly tested before being approved for use; to help in reducing the risks associated with it to a low percentage.

Since the main reason sustainable agriculture is being advocated by several environmental bodies is because of global warming, a term was introduced to help in the description of the process of reducing the concentration of carbon dioxide, which is one of the main greenhouse gases that promote global warming. Carbon sequestration refers to the use of agricultural and forestry practices to help in removing carbon dioxide from the atmosphere. With carbon sequestration, global climate could be adjusted by enhancing carbon storage in trees and soils, hence preserving existing tree and soil carbon (Chhipa, H. 2017).

Carrying capacity: this is a term used in sustainable agriculture to refer to the maximum population size that can be supported in terms of food availability indefinitely without there being the risk of damaging the environment to an extent that it would be good for use by future generations. Most environmental bodies determine the earth's capacity using both natural constraints and the choices human beings make based on economics, environment, culture, and topography. However, in most cases, the carrying capacity is mostly an approximation of what it should be, since it is quite a challenge to determine the exact carrying capacity.

Community-supported agriculture (CSA): this is an important term in sustainable agriculture for it supports the growth of sustainable contaminant-free agriculture by allowing members of a community to willingly offer their support to activities on a farm, making the farmland legally a community's farm. Within a community farm, both the growers and the consumers provide mutual support and share the risks and the benefits which are encountered in food production. Generally, community-supported agriculture is all about the relationship between the farm's shareholders and the farmers. The shareholders cater for the cost of all farm activities and the payment of farmers' salaries, and they get their profits by receiving shares in the farm harvest throughout the growing season. Community-supported agriculture promotes direct marketing by allowing farmers to sell their harvest directly to community members, providing the farmers with working capital in advance and also getting better prices for their crops, an activity that promotes the farmers' financial security.

Conservation buffer strips: this is a concept used in sustainable contaminant-free agriculture whereby strips of land are maintained in permanent vegetation, serving as an intercept to pollutants and soil erosion. Such strips help enhance wildlife habitats, improve water quality and enrich farmlands. There are several types of conservation buffer strips and including contour buffer strips, filter strips, riparian forest buffers, field borders, windbreaks, hedgerows, grassed waterways, and alley cropping.

Conservation tillage: this term has a close meaning to conservation buffer strips with the main difference being that residue is left on the soil surface, substantially minimizing the effects of soil erosion mainly from water and wind. Tillage does also reduce nutrient loss by releasing nutrients in the soil upon the decay of the residue cover. With conservation tillage, the soil is to be left undisturbed until harvest time except during nutrient amendment.

In sustainable contaminant-free agriculture, it is necessary to track the use of natural resources and the waste produced after their use. The ecological footprint is a term used to refer to the measure of how much land and water is needed for agricultural production and the disposal of the waste produced during agricultural production. Carbon footprint, as the name suggests, is an ecological footprint that is used to represent the effect agricultural activities have on climate change in terms of the greenhouse gases produced.

Eco-label: this is a term used to refer to a seal indicating if a certain agricultural product has met a set of environmental or social standards. According to sustainable contaminant-free agriculture, the agricultural products produced have to be of a specific quality, and if it is not met, the production is said to have failed in promoting sustainable agriculture. The labels are a way of communicating the pride in sustainable agriculture in different nations. An eco-label is assigned to a specific agricultural product by a third party based on outlined environmental leadership criteria.

Environmental indicator: this is a term used to refer to information about the environment, highlighting the effects that human activities have on the environment and the emerging problems facing the environment, and also drawing attention to the benefits which have been gained from the implementation of the current policies (Chhipa, H. 2017). There are different types of environmental indicators in sustainable agriculture. An agri-environmental indicator communicates information about the change in the state of environmental resources used or resources that have been affected by agricultural activities. An example of the resources monitored by agri-environmental indicators includes soil quality, water quality, agroecosystem biodiversity, climate change, farm resource management, and farm production efficiency. Externalities are used to refer to the benefits and costs that are not included in the market prices of agricultural products. The quantification of externalities in sustainable agriculture is crucial for it helps in the determination of agricultural production level and in the formulation of policy decisions that are aimed at promoting the growth of sustainable agriculture.

Farmland protection: this is a crucial term for sustainable agriculture for it advocates for the protection of irreplaceable farmlands, and wildlife habitat from urban sprawl and rural subdivisions. Without farmland protection, most farmlands would end up being destroyed, affecting the growth of sustainable agriculture in different parts of the world. The destruction of wildlife habitat would also affect sustainable agriculture negatively by causing ecological imbalances. The different actions that have been drawn out to help in farmland protection include the purchase of agricultural conservation easement programs that aim at making the exercise of farmland protection simple; establishment of special agricultural districts where commercial agriculture is to be practiced ensuring that property tax is well taken care of; farm-friendly zoning ordinances; and comprehensive land use planning, a move aimed at discouraging farmland grabbers from grabbing the unused agricultural lands.

Sustainable contaminant-free agriculture does also focus on livestock rearing. Grass-based farming is a term used to refer to the reliance on pasture to supply the needed protein and energy requirements of livestock. Grazing is aimed to replace close confinement feeding of animals, giving domesticated animals the freedom of movement. For grazing to be a success, the farmer ought to focus on pasture plant and soil management, ensuring that proper stocking density is maintained to avoid the negative impacts of overgrazing. With grass-based farming, farm inputs such as diesel, fertilizers and purchased feeds are minimized which helps in improving a healthy farm. Limited tillage is made possible through the use of remaining pastures which store carbon in the soil hence leading to the accumulation of organic soil matter, which conserves the soil. Good quality plant and animal genetics are selected to ensure that the final agricultural products gotten from them are of the best quality. Grass-based farming leads to the existence of two types of animals. Grass-fed animals refer to animals that have been raised by being fed purely on grass and hay, whereas, grass-finished animals refer to animals that have grown fast enough on pasture but not purely on grass throughout their life.

Holistic management: this is a term used to refer to the farm system approach to resource management that involves financial planning, land planning, grazing planning, and biological monitoring. The main benefit of holistic management in sustainable contaminant-free agriculture is that it promotes soil health, helps in curbing soil erosion within the farm, improves biodiversity by ensuring ecological balance is maintained and enhances agricultural productivity since it aims at working with Nature forces.

The main reason for the existence of sustainable contaminant-free agriculture is to ensure that both the food system and farming system are viewed as a whole system. Integrated food and farming system allows farms and food production systems to be integrated into one, a move that allows for the efficient use of natural, economic, and social resources (Chhipa, H. 2017). Pests are a big problem in all agricultural practices, be it the traditional agricultural methods or the new sustainable contaminant-free agriculture. Integrated pest management makes it possible to address the pest problem through an ecological approach that utilizes the knowledge of crop and pest relationships, allowing for the establishment of acceptable economic thresholds for pest populations ensuring that they do not get to a percentage of threatening to be a potential problem in the practice of sustainable agriculture. Biointensive integrated pest management focuses on the use of preventive tactics and biological controls that aim at maintaining the pest population within acceptable limits. Biological control, as the name suggests, is the use of specially chosen living organisms that are predators of a particular pest.

Low-input agriculture is a major practice in sustainable agriculture for it helps in optimizing the management and usage of internal production inputs and also minimizes the use of farm inputs to lower the farm's production costs and further avoid the pollution of surface and groundwater; a move that reduces pesticide remains in food being grown in a farm. In general, low-input agriculture focuses on reducing a farmer's overall risk hence increasing both the short-term and long-term profitability of the farm's produce. Nutrient management is used to refer to the management of the amount, source, addition, form, and timing of the application of soil amendments to ensure adequate soil fertility that boosts plant product quality and reduces the risk of environmental degradation.

10.3 UNITED NATIONS ENVIRONMENT PROGRAM (UNEP)

The United Nations Environment Program is an environmental body that is responsible for handling responses to environmental issues within the United Nations Systems. The United Nations system is a body made of several member states from all over the world that have common beliefs about several life issues. UNEP was first founded in 1972 after the United Nations meeting on the Human environment. The main reason for the establishment of UNEP was to ensure that there is leadership, delivery in science, and development of solutions to a wide range of issues concerning the environment such as climate change, management of marine and terrestrial ecosystems, and green economic development among other issues.

The UNEP organization is responsible for addressing several issues some of which include, the formulation of international environmental agreements that ensure that all member states follow similar environment conservation policies; green economic development movements that ensure that all member states have common development projects that aim at improving the condition of the environment at large; publishing and promotion of environmental science whose main purpose is to help in improving a healthy environment across the globe; and helping national governments of their respective member states in achieving their environmental targets. Since the establishment of the United Nations Environment Program, it has been serving as a mother or a backbone to several environment-friendly organizations. UNEP hosts the secretariats of several environmental research bodies such as the Convention on Biological Diversity (CBD), the Minamata Convention on Mercury (MCM), and the Convention on International Trade in Endangered Species of Wild Fauna and Flora (CITES). UNEP does also work hand-in-hand with other environmental bodies to help in the establishment of sub-bodies that help in promoting a clean environment. For instance, UNEP in association with the World Meteorological Organization formed the Intergovernmental Panel on Climate Change whose aim is on monitoring climate change, and research the best ways how to address the issue of climate change.

The establishment of UNEP faced several challenges with the main challenge being that of convincing developing nations to take part in environmental conservation. Most developing nations believed that they had better things to focus on like developing their nations, instead of having to focus on environmental protection. However, with the persuasion of the co-founders of UNEP, most developing nations and developed nations ended up being part of the UNEP organization. The governance of UNEP is made up of several positions. The first position is that of the executive director who is responsible for governing the whole UNEP organization. The second position is the Environment Assembly which is made of up a hundred and ninety-three members who meet once every two years (Chhipa, H. 2017).

Divisions of the UNEP

Generally, the UNEP organization is divided into eight divisions. The first division is the science division which aims at providing scientifically credible environmental assessments and information for sustainable environmental development. It is through the science division that sustainable contaminant-free agriculture is most advocated for. The science division does also report on the state of the environment from a global perspective. It assesses the environmental policies established in different member states weighing if they are eligible or not. It provides warning of the emerging environmental threats giving a summary of the possible quick solutions that might help in curbing such threats. The UNEP recently established a 2030 vision, whose progress is monitored and reported on by the science division.

The second division of the UNEP organization is the Policy and Program Division. As the name suggests, it is responsible for the making of policies that govern environmental protection. It does also have the responsibility of ensuring that all the other divisions of the UNEP are well coordinated and working effectively. The economy division has the role of assisting large businesses in their goal of being more environmentally conscious. It is further categorized into three groups. The first group is the chemicals and health category. In terms of sustainable contaminant-free agriculture, the chemicals and health category help in ensuring that the farm inputs are environmentally friendly and cause minimal risk to the health of the farm and the consumers of the farm's produce. The second group is the energy and climate category. The energy and climate category contributes to sustainable agriculture by innovating agricultural methods which aid in reducing the total energy consumption in agricultural productions and reducing the emission of greenhouse gases within the farm that could contribute to climate change. The third group is the resources and markets category. As the name suggests, the resources and markets category contribute to sustainable contaminant-free agriculture by ensuring that the farm resources are efficiently utilized and by promoting direct marketing of the farm's produce, eliminating the third-party individual.

UNEP is an organization that is well known for good interworking relations with other environmental bodies. The governance affairs office division is responsible for overseeing the engagement of member states and other relevant groups in using the already existing UNEP work to help in making the environment healthier. The office does mostly serve the main governing body of UNEP, the Committee of Permanent Representatives, and oversees their regular meetings. The main benefit of the government affairs office division is that it helps in strengthening the visibility, authority, and impact of the assembly as an authoritative voice on the environment.

Law division as the name suggests helps in the development of environmental law ensuring that the environment is well taken care of. Members of this division work hand-in-hand with member states to help in eliminating environmental crime ensuring that the set international environmental commitments are met. The main purpose of the law division is to improve the cooperation between lawmakers around the world who solely focus on the formulation of environmental laws. In terms of sustainable agriculture, the law division makes it easy for rules governing sustainable contaminant-free agriculture easy to follow.

The communication division focuses on the formulation and distribution of UNEP's messages to its member states and also to other environmental bodies with which it is working at the time. This division ensures that the messages are delivered to governments and individuals either through digital or traditional media channels. UNEP uses this division to help in communicating new practices in sustainable agriculture (Chhipa, H. 2017).

The last division of UNEP is the corporate service division. It is responsible for handling UNEP's corporate interests such as the management of financial risk and also the exposure of such risks.

UNEP Activities that Boost Sustainable Contaminant-free Agriculture

The UNEP organization outlined several different activities which go hand-in-hand with sustainable contaminant-free agriculture. The first activity is related to climate change. Climate change is a big issue in sustainable agriculture. Most projects in sustainable agriculture are aimed at minimizing the effects of climate change on agricultural activities. The second UNEP activity is disasters and conflict management. UNEP has taken the responsibility of investigating the influence of emergencies and natural disasters on human health and most specifically on sustainable contaminant-free agriculture. Through this activity, UNEP helps in reducing the possible causes of disasters by controlling the ecological balance. The main reason UNEP activities associated with disasters and conflicts are important in sustainable agriculture is that they help member countries in the implementation of policies that prevent serious damage caused by disasters. Preventing such disasters promotes the growth of sustainable agriculture in different parts of the world.

Ecosystem management is the third UNEP activity. UNEP ensures that the ecosystem is well managed, a move that helps in sustainable agriculture by ensuring that all the components that make up the environment are in perfect condition and their interactions with each other are well coordinated without leading to an ecosystem imbalance. Environmental governance is a UNEP activity that focuses on ensuring that environmental practices, more so agricultural environmental activities are done in an appropriate that does not hurt the environment. Chemicals and waste management is a UNEP activity that ensures that all the practices in sustainable agriculture ensure minimal use of chemicals and that the waste produced during agricultural waste production.

Resource efficiency is another UNEP activity that focuses on ensuring that both environmental and non-environmental resources are utilized well. In sustainable contaminant-free agriculture, resource efficiency will ensure that the farm resources are not overused or underused. UNEP does also have another activity that majorly focuses on promoting sustainable contaminant-free agriculture. The environment under review, as the name suggests, is a UNEP activity that ensures that UNEP supplies data and information on the traits of the global environment to governments and other non-governmental organizations to help them in implementing sustainable development goals such as sustainable agriculture. The information being shared is always from the most recent research that has been done and approved within the UNEP organization. This ensures that the sustainable agriculture policies that are being implemented follow reliable information effectively.

10.4 GLOBAL ALLIANCE ON HEALTH AND POLLUTION (GAHP)

The Global Alliance on Health and Pollution is an environmental body that is made up of a network of international and national level agencies, whose commitment is to a collaborative and multi-sectoral approach in addressing the global pollution menace and the resulting negative impacts on human health and environmental health at large and also the negative impact on the economics of different parts of the world. The GAHP is made up of more than sixty members. With the help of extensive research, GAHP has managed to include almost all forms of pollution that have ever existed in the world in its records; a move which has helped in it achieving several sustainable development goals, with sustainable contaminant-free agriculture being part of them. GAHP organization has had the privilege of launching health and pollution implementation plans in almost ten countries across the world.

The main goals of GAHP contribute a lot to sustainable agriculture for it focuses on reducing death and illnesses that have been caused by all forms of harmful and toxic pollution, including air, water, soil, and chemical wastes which are the main challenge, especially in developing countries. Apart from being made up of more than sixty members, GAHP does have several observers whose main job is pushing for the provision of resources that will aid in coming up with solutions to pollution problems.

The establishment of GAHP was brought about by the need of having an international and also national level body that would be able to acknowledge that a collaborative, multi-stakeholder and multi-sectoral approach would be paramount in dealing with the global pollution crisis which leads to a negative health and economic impact not only to the developing countries but also to the developed nations since pollution is all over the globe. In the year two thousand and twelve, the foundation of GAHP was laid by

Pure Earth organization as it initiated a union with representatives from the World Bank, UNEP, UNDP, UNIDO, Asian Development Bank, the European Commission, and Ministries of Environment and Health of several countries; a union that would help in the formulation of strategies aimed at addressing pollution and health issues caused by pollution. In the year two thousand and nineteen, GAHP was officially included in addressing global pollution and health issues.

Role of GAHP in Sustainable Contaminant-free Agriculture

The GAHP organization does play a crucial role in the growth of sustainable agriculture by focusing on two main areas. The first one is the advocacy and raising awareness on global pollution and its effects on health and in this case, particularly agricultural health. The second role is country-specific support in addressing global pollution. Country-specific support in addressing the global pollution crisis is made possible by GAHP building public, technical, and financial support systems that aim at promoting scientific research on the causes of global pollution and possible solutions to the health issues that are experienced due to pollution. In most cases, GAHP guides members in a journey of prioritizing the pollution problem and addressing it from its root cause through the implementation of health and pollution action plans.

GAHP organization is also known for working hand in hand with other environmentally friendly organizations to boost sustainable agriculture through the reduction of global pollution. For instance, in the year two thousand and seventeen, GAHP in collaboration with the Lancet organization published an article on pollution and health highlighting possible ways that can be used in addressing full health and economic costs of air, water, and soil pollution. The published research information was based on the analysis of existing and emerging data, which helped in revealing the severe and unreported contribution different forms of pollution have to global diseases, which keep on emerging daily. The publication did also highlight the deaths caused by pollution on a global level.

Global pollution is known to affect sustainable agriculture through effects such as climate change, and contaminated irrigation water which ends up contaminating agricultural products, among others. Furthermore, agricultural activities are also known to contribute to global pollution through the excessive use of farm inputs such as the likes of pesticides, herbicides, and fertilizers. According to research by GAHP, agricultural pollution contributes to the three main forms of pollution. Excessive use of fertilizers does lead to possible loss of soil fertility which in a way is soil pollution. Herbicides being used to control weeds could flow down to nearby water bodies through run-off rainwater leading to water pollution. The fertilizers used in farm production, if used in plenty, could lead to the release of greenhouse gases which contribute to air pollution.

Generally, the GAHP organization helps in providing information that will help sustainable contaminant-free agriculture to stick to ways that reduce global pollution and also reduce the global burden of disease.

10.5 EUROPEAN ENVIRONMENT AGENCY (EEA)

The European Environment Agency is an environmental body made up of member states of the European Union, and it provides independent information on the environment. In terms of sustainable contaminant-free agriculture, the EEA helps in providing relevant information that could help in the growth of sustainable agriculture across its member states. Its main goal is to ensure that all its members are focused on the development of environmental laws some of which promote sustainable agriculture, the implementation and evaluation of the formulated laws, and also in ensuring that the public is made aware of them and that it does strictly adhere to them.

The European Environment Agency was established by the European Economic Community (EEC) in the year 1990, and it later became fully functional in the year 1994. The structure of the EEA organization is made up of three divisions. The first division comprises representatives of the governments from all the EEA member states. The second division is comprised of mainly two internationally recognized scientists who are selected by the European Parliament and assisted by the scientific committee. The scientists are responsible for overseeing the agency's research projects. The third division is that of the executive director who serves a term of five years, and he or she is responsible for overseeing the activities of the entire agency. The EEA is made up of thirty-two member countries and the other six that are cooperating countries. Cooperating countries are not members of the Europe Nations body hence the name cooperating countries.

Role of EEA on Sustainable Contaminant-free Agriculture

When it comes to sustainable contaminant-free agriculture, the European Environment Agency has the responsibility of focusing on the production of assessments based on quality-assured data on a wide range of issues from biodiversity (ensuring that biodiversity is taken care of during sustainable agriculture activities), air quality (improving the quality of farm produce through the reduction of carbon dioxide concentration in the atmosphere for it does contaminate the soil if it is in high concentration within the atmosphere), and the journey to climate change (which highlights the possible causes of climate change within a farm practicing sustainable agriculture). The results of the EEA assessments help in tracking the progress in sustainable agriculture, hence indicating areas where additional efforts are needed to boost the growth of sustainable agriculture across the member states of the EEA (Chhipa, H. 2017). Results from the regular EEA assignments are frequently posted on its main websites and several information platforms such as Biodiversity Information Systems for Europe and the ClimateADAPT platform; providing easy and fast access to information on sustainable agriculture. The ClimateADAPT platform majorly focuses on sharing EEA assessment result information on expected climatic conditions, the different regions and sectors that are vulnerable to climate change, adaptation case studies, adaptation options, and the adaptation planning tools to be used in facing the effects that could be caused by the climate change.

The EEA has an information system referred to as the European Nature Information System (EUNIS). The EUNIS is a public database that contains information about plant and animal species, their habitats, and protected sites across Europe. Information on both plant and animal species is quite detailed allowing farmers who practice sustainable contaminant-free agriculture to gain more knowledge on the best species of plants and animals to take care of within their farms based on the information on their habitat conditions.

Just like most environmentally friendly organizations have sub-bodies, the EEA works in partnership with the European Environment Information and Observation Network (Eionet), which is a network of the EEA and its member countries. The Eionet network is responsible for developing the network and coordinating EEA activities at a specific level by working closely with the national focal points (NFPs), which are the environmental ministries in respective member countries of the EEA organization.

To promote trust in the EEA organization, the EEA gives full access to its administrative and budgetary documents in its documents register.

In conclusion, sustainable contaminant-free agriculture is the future of the agricultural world across the globe. It ought to be advocated for not only by environmentally friendly organizations but also by governments and their citizens. With sustainable contaminant-free agriculture, the future of agriculture is bright since it aims at facing all the challenges head-on and overcoming them.

REFERENCES

1. Akeel, A., & Jahan, A. . (2020). Role of cobalt in plants: its stress and alleviation. *In Contaminants in agriculture*, (pp. 339-357). Springer, Cham.
2. Ansari, A. A., Naeem, M., & Gill, S. S. . (2018). Contaminants in agriculture: Threat to soil health and productivity. . *Soil Science Society of America Journal*, 68, 1945-1962.
3. Bastida F, Moreno JL, Hernandez T, & Garcia C . (2006). Microbiological degradation index of soils in a semiarid climate. *Soil Biology and Biochemistry*, 38(12): 3463-3473.
4. Boxall, A. B., Hardy, A., Beulke, S., Boucard, T., Burgin, L., Falloon, P. D., ... & Williams, R. J. (2009). Impacts of climate change on indirect human exposure to pathogens and chemicals from agriculture. *Environmental health perspectives*, 117(4), 508-514.
5. Breure AM, Mulder CH, Rutgers M, Schouten T, & De Zwart D, et al. . (2004). biological indicator for soil quality. In: proceedings from an OECD expert meeting Rome, Italy, March 2003: agricultural impacts on soil erosion and soil biodiversity. *Developing indicators for policy analysis*, pp. 485-494.
6. Brodie, J. E., Kroon, F. J., Schaffelke, B., Wolanski, E. C., Lewis, S. E., Devlin, M. J., ... & Davis, A. M. . (2012). Terrestrial pollutant runoff to the Great Barrier Reef: an update of issues, priorities and management responses. *Marine pollution bulletin*, 65(4-9), 81-100.
7. Chauhan A, M. B. (2015). Soil Health - An Issue of Concern for

Environment and Agriculture. *J Bioremed Biodeg*, 6: 286.

8. Debnath, B., & Ahammed, G. J. (2020). Effect of acid rain on plant growth and development: physiological and molecular interventions. In Contaminants in Agriculture. (pp. 103-114).
9. Fairbairn, D. J., Arnold, W. A., Barber, B. L., Kaufenberg, E. F., Koskinen, W. C., Novak, P. J., ... & Swackhamer, D. L. . (2016). Contaminants of emerging concern: mass balance and comparison of wastewater effluent and upstream sources in a mixed-use watershed. . *Environmental Science & Technology*, 50(1), 36-45.
10. Food and Agriculture Organization of the United Nations. (2018). *Soil pollution: a hidden reality.* London: Food & Agriculture Org. pp.30 - 156.
11. Fouché, O., Lasagna, M., & Danert, K. . (2019). Groundwater under threat from diffuse contaminants: improving on-site sanitation, agriculture and water supply practices. *Environmental Science and Pollution Research*, 26(3), 2023-2025.
12. Fries W, Fried J, Platzer K, Horak O, Gerzabek MH . (2006). Remediation of contaminated agricultural soils near a former Pb/Zn smelter in Austria: Batch, pot and field experiments. *Environmental Pollution*, 144 (1): 40-50.
13. Giannakis, E., Kushta, J., Giannadaki, D., Georgiou, G. K., Bruggeman, A., & Lelieveld, J. (2019). Exploring the economy-wide effects of agriculture on air quality and health: Evidence from Europe. . *Science of The Total Environment*, 663, 889-900.
14. Goyal, D., Yadav, A., Prasad, M., Singh, T. B., Shrivastav, P., Ali, A., ... & Mishra, S. (2020). Effect of heavy metals on plant growth: an overview. . *Contaminants in agriculture*, 79-101.
15. Graham Merrington, Dr. Linton Winder Nfa, R. Parkinson, Mark Redman, & L. Winder. (2002). *Agricultural Pollution: Environmental Problems and Practical Solutions.* Florida: CRC Press, pp22 - 264.
16. Hai, D. M., Qiu, X., Xu, H., Honda, M., Yabe, M., Kadokami, K., ... & Oshima, Y. (2017). Contaminants in liquid organic fertilizers used for agriculture in Japan. *Bulletin of environmental contamination and toxicology,*, 99(1), 131-137.
17. Heathwaite, A. L., Quinn, P. F., & Hewett, C. J. M. (2005). Modelling and managing critical source areas of diffuse pollution from agricultural land using flow connectivity simulation. *Journal of Hydrology,* , 304(1-4), 446-461.

18. Huff Chester, A., Gordon, C., Hartmann, H. A., Bartell, S. E., Ansah, E., Yan, T., ... & Schoenfuss, H. L. . (2022). Contaminants of emerging concern in the Lower Volta River, Ghana, West Africa: The agriculture, aquaculture, and urban development nexus. *Environmental Toxicology and Chemistry*, , 41(2), 369-381.
19. Ioannidou, V., & Stefanakis, A. I. . (2020). The use of constructed wetlands to mitigate pollution from agricultural runoff. . *In Contaminants in Agriculture*, (pp. 233-246). Springer, Cham.
20. Joint FAO/WHO Expert Committee on Food Additives. Meeting, & World Health Organization. (2006). Safety evaluation of certain contaminants in food (Vol. 82). *Food & Agriculture Org..*, .
21. Joint, F. A. (2011). *Safety evaluation of certain contaminants in food: prepared by the Seventy-second meeting of the Joint FAO/WHO Expert Committee on Food Additives (JECFA)*, World Health Organization.
22. Karić, N., Maia, A. S., Teodorović, A., Atanasova, N., Langergraber, G., Crini, G., ... & Đolić, M. . (2022). Bio-waste valorisation: Agricultural wastes as biosorbents for removal of (in) organic pollutants in wastewater treatment. *Chemical Engineering Journal Advances,* , 9, 100239.
23. Kaushal, S. S., Mayer, P. M., Vidon, P. G., Smith, R. M., Pennino, M. J., Newcomer, T. A., ... & Belt, K. T. (2014). Land use and climate variability amplify carbon, nutrient, and contaminant pulses: a review with management implications. *JAWRA Journal of the American Water Resources Association*, 50(3), 585-614.
24. Kroeze, C., Gabbert, S., Hofstra, N., Koelmans, A. A., Li, A., Löhr, A., ... & van Wijnen, J. (2016). Global modelling of surface water quality: a multi-pollutant approach.. *Current opinion in environmental sustainability*, 23, 35-45.
25. Kumar, V., Kumar, R., Singh, J., & Kumar, P. (Eds.). (2019). Contaminants in Agriculture and Environment: Health Risks and Remediation (Vol. 1). . *Agro Environ Media, Publication Cell of AESA, Agriculture and Environmental Science Academy.*
26. Lamastra, L., Suciu, N. A., & Trevisan, M. (2018). Sewage sludge for sustainable agriculture: contaminants' contents and potential use as fertilizer. *Chemical and Biological Technologies in Agriculture,*, 5(1), 1-6.
27. Li, W., Li, X., Tong, J., Xiong, W., Zhu, Z., Gao, X., ... & Liang, J.

(2022). Effects of environmental and anthropogenic factors on the distribution and abundance of microplastics in freshwater ecosystems. *Science of The Total Environment*, 159030.

28. M. Naeem, Abid Ali Ansari, & Sarvajeet Singh Gill. (2020). *Contaminants in Agriculture: Sources, Impacts and Management.* Berlin: Springer Nature, pp.25 - 446.
29. Marella, T. K., Saxena, A., Tiwari, A., Datta, A., & Dixit, S. (2022). Treating agricultural non-point source pollutants using periphyton biofilms and biomass volarization. *Journal of Environmental Management*, 301, 113869.
30. Margenat, A., Matamoros, V., Díez, S., Cañameras, N., Comas, J., & Bayona, J. M. (2017). Occurrence of chemical contaminants in peri-urban agricultural irrigation waters and assessment of their phytotoxicity and crop productivity. *Science of the Total Environment*, 599, 1140-1148.
31. Margenat, A., Matamoros, V., Díez, S., Cañameras, N., Comas, J., & Bayona, J. M. . (2019). Occurrence and human health implications of chemical contaminants in vegetables grown in peri-urban agriculture. *Environment International*, 124, 49-57.
32. Mateo-Sagasta, J. Z. (2017). *Water pollution from agriculture: a global review. Executive summary.* Rome: Ecosystems (WLE), 112-140.
33. Muñoz, I., Gómez-Ramos, M. J., Agüera, A., Fernández-Alba, A. R., García-Reyes, J. F., & Molina-Díaz, A. (2009). Chemical evaluation of contaminants in wastewater effluents and the environmental risk of reusing effluents in agriculture. *Trends in Analytical Chemistry*, 28(6), 676-694.
34. Neal, C., Jarvie, H. P., Withers, P. J., Whitton, B. A., & Neal, M. (2010). The strategic significance of wastewater sources to pollutant phosphorus levels in English rivers and to environmental management for rural, agricultural and urban catchments. *Science of the Total Environment*, 408(7), 1485-1500.
35. Prasad, M., Ranjan, R., Ali, A., Goyal, D., Yadav, A., Singh, T. B., ... & Dantu, P. K. (2020). Efficient transformation of agricultural waste in India. *In Contaminants in Agriculture*, (pp. 271-287). Springer, Cham.
36. Rashmi, I., Roy, T., Kartika, K. S., Pal, R., Coumar, V., Kala, S., & Shinoji, K. C. (2020). Organic and inorganic fertilizer contaminants in agriculture: Impact on soil and water resources. *In Contaminants in agriculture*, (pp. 3-41).

37. Rigby, H., Dowding, A., Fernandes, A., Humphries, D., Jones, N. R., Lake, I., ... & Smith, S. R. (2021). Concentrations of organic contaminants in industrial and municipal bioresources recycled in agriculture in the UK. *Science of the Total Environment*, 765, 142787.

38. Roccaro, P., & Vagliasindi, F. (2014). Risk assessment of the use of biosolids containing emerging organic contaminants in agriculture. *Chemical Engineering*, 37.

39. Shrivastav, P., Prasad, M., Singh, T. B., Yadav, A., Goyal, D., Ali, A., & Dantu, P. K. (2020). Role of nutrients in plant growth and development. *Contaminants in agriculture*, (pp. 43-59). Springer, Cham.

40. Siddiqa, A., & Faisal, M. (2020). Heavy metals: source, toxicity mechanisms, health effects, nanotoxicology and their bioremediation. *In Contaminants in Agriculture*, (pp. 117-141). Springer, Cham.

41. Singh, P., Raj, A., & Yadav, B. (2022). Impacts of Agriculture-Based Contaminants on Groundwater Quality. *In Sustainability of Water Resources*, (pp. 249-261). Springer, Cham.

42. Singh, P., Raj, A., & Yadav, B. (2022). Impacts of Agriculture-Based Contaminants on Groundwater Quality. *In Sustainability of Water Resources*, (pp. 249-261). Springer, Cham.

43. Singh, T. B., Ali, A., Prasad, M., Yadav, A., Shrivastav, P., Goyal, D., & Dantu, P. K. (2020). . (2020). Role of organic fertilizers in improving soil fertility. . *In Contaminants in agriculture*, (pp. 61-77). Springer, Cham.

44. Stoate, C., Baldi, A., Beja, P., Boatman, N. D., Herzon, I., Van Doorn, A., ... & Ramwell, C. . (2009). Ecological impacts of early 21st century agricultural change in Europe–a review. *Journal of environmental management*, 91(1), 22-46.

45. Sunyer-Caldú, A., Sepúlveda-Ruiz, P., Salgot, M., Folch-Sánchez, M., Barcelo, D., & Diaz-Cruz, M. S. (2022). Reclaimed water in agriculture: a plot-scale study assessing crop uptake of emerging contaminants and pathogens. *Journal of Environmental Chemical Engineering*, 108831.

46. Varjani, S., & Upasani, V. N. . (2019). Influence of abiotic factors, natural attenuation, bioaugmentation and nutrient supplementation on bioremediation of petroleum crude contaminated agricultural soil. . *Journal of environmental management*, 245, 358-366.

47. Wortman, S. E., & Lovell, S. T. (2013). Environmental challenges

threatening the growth of urban agriculture in the United States. *Journal of environmental quality*, 42(5), 1283-1294.

48. Zhang, L., Yan, C., Guo, Q., Zhang, J., & Ruiz-Menjivar, J. (2018). The impact of agricultural chemical inputs on environment: global evidence from informetrics analysis and visualization. *International Journal of low-Carbon technologies*, 13(4), 338-352.
49. Zhang, X., Chen, L., & Shen, Z. . (2021). Impacts of rapid urbanization on characteristics, sources and variation of fecal coliform at watershed scale. . *Journal of Environmental Management*, 286, 112195.

INDEX

A

Abiotic 22
Aerobic 134
Agrarianism 196
Agricultural pesticides 15
Agricultural pollution 2, 181
Agricultural process 77
Agriculture 22, 158
Agritourism 195
Agrochemicals 38, 150
Agroecosystems 157
Agroforestry 195
Airborne 160
Air pollution 2
Air sparging 141
Algae 96
Alzheimer's disease 50
Ammonia 6, 89
Ammonia volatilization 33
Amyotrophic lateral sclerosis (ALS) 159
Anaerobic 134
Anaerobic organisms 5
Analgesics 33
Analytic 84
Anthropogenic eutrophication 131
Antibiotic 152
Antiepileptics 33
Antimicrobial 31
Antimicrobial-resistant bacteria 39
Antimicrobials 31, 45, 82
Antiretrovirals 33
Aquaculture 44, 155, 158
Aquaculture systems 44
Aquaculture waters 163
Aquatic animals 88
Aquatic life 18, 183
Aquatic organisms 101
Aquatic plants 90

Arable land 23
Asthma 15, 116
Asymmetry 167
Atmosphere 162
Auditors 117
Authorities 171
Authorize 173

B

Bacteria 5
Bioaccumulation 54, 64
Bio-augmentation 134
Bioaugmentation 146
Bioavailability 11, 135
Biocells 143
Biodegradable 128
Biodegradation 135
Biodegradation process 35
Biodiversity 150, 176
Bioleaching 134, 140
Biological amplification 64
Biological degradation 131
Biopiling 139
Bioreactors 140
Bioremediation 134
Bioremediation processes 139
Bio-sparging 146
Biotic 22
Biotic populations 10
Bioventing 135
Biovolatilization 134
Blood 84
Burn 177
Byproducts 32

C

Cadmium (Cd) 10
Carbohydrates 50
Carbon dioxide 6
Carcinogenic 35
Carnivores 55
Catalyzing reactions 69
Cereals 22
Chemical pesticides 44
Chromium 28
Climate 98
Climatic 128
Collaboration 93
Comas 12
Combination 128
Comfortability 122
Commodity prices 88
Community 35
Composting 140, 143
Conceptualization 174
Consumers 110, 112, 121
Containers 34
Contamination 2
Contemporaneous single 172
Contemporary technologies 178
Corncobs 143
Cow burps 161
Cropping 44
Crop production 44
Cryptosporidium 102
Cultivated crops 8
Customers 114
Cyanide 62
Cytotoxicity 51

D

Dangerous 128
Dechlorination 135
Decomposition process 19
Deforestation 8, 179
Degradation 2
Delineation 173
Demographic 157

Demolition 6
Denitrification 106
Denitrification processes 165
Density 20
Depletion 104
Detection 84
Devastating 89
Diagnosis 150
Dioxins 130
Disastrous consequences 44
Distributors 194
Dolphins 64
Donors 26
Drainage 172
Drastic measures 128
Dredged sediments 24
Drippers 34
Drug remnants 150
Dust 157
Dwelling 62

E

Economic factors 172
Economic impact 116
Economic worth 105
Ecosystem 161
Eeutrophication 33
Electricity 82
Electrokinetic remediation 141
Electronic 6
Electronic device 83
Electronic wastes 83
Electrostatic filtering 14
Employees 121, 122
Endanger 25
Endocrine system 50
Enteric fermentation 4, 161
Environmentalists 68
Epilepsy 12
Erosion 19, 171, 180
Eutrophication 6, 101, 166, 170
Evaporation 34, 96
Excavation 38
Execution 130
Expenditures 35
Ex situ biological treatment 139
Ex situ bioremediation 142
Extension system 154

F

Farmers 45
Farm pollution 157
Farms 20
Fast erosion 130
Fauna 129
Feed parameters 162
Fermentation 161
Fertility 157
Fertilization 15
Fertilizer 2
Fishing zones 96
Flagging economics 44
Flora 129
Fluorine 28
Food and Drug Administration (FDA) 110
Foodborne diseases 121
Forage crops 22
Fungicides 24, 97

G

Gamma isomer 41
Gastrointestinal disorders 116
Gene manipulations 112
Genetic mutations 111
Geological erosion 130
Geology 172, 183

Global Environmental Fund (GEF) 26
Global market 23
Gluten 50
Glyphosate 50
GMOs 111
Grazing 2
Greenhouse gas emissions 161
Greenhouse gases 117
Groundwater 2
Groundwater salinization 44
Gypsum 128

H

Harmful elements 14
Harmless molecules 183
Harvesting 176
Harvests 36
Hazardous 62, 69
Hazardous chemicals 64
Hazardous compounds 28, 170
Hazardous gases 5
Hazardous metals 31
Hazardous risks 109
Heavy metals 11
Herbicides 22, 97, 156, 195
Heterogeneous 128
Homogeneity 139, 141
Hormones 8, 33
Horticultural gardens 37
Hyacinths 90
Hydrocarbons 134
Hydrogen sulfide 5
Hydrophilic pollutants 32

I

Ignorance 157
Implementation 49
Inadequate agricultural techniques 130
Inadequate grass cover 184
Inadequate management 19
Incentive program 113
Incentives 112
Increase 23
Infection 173
Infectious 5
Infiltration 184
Inherent quality 19
Innovative devices 180
Inorganic 22
Inorganic fertilizers 109
Insecticides 17, 22, 97, 134
In-situ bioremediation 144
Instructors 171
Intensification 24
Intensive farming 175
Interactions 130
Intercropping 103
International Organization for Standardization (ISO) 118
Intrinsic bioremediation 144
Investigation 172
Irrigation 22, 77, 177
Isotopic 162

L

Lagoons 89
Lakes 3
Landfills 129
Landscape-level pollution 2
Lawsuits 122
Leaders 171
Legislation 119, 171
Liability 120
Lipophilic pollutants 32
Livestock 44, 155

Livestock production 8

M

Magnification 64
Malathion 49
Mammals 17
Manure 183
Margins 101
Marine animals 92
Marketing 164
Mass 88
Matrix 39
Maximum residual limits authorized. (MRL) 10
Medication 89
Mercury 62
Metabolic capacities 69
Metabolic cycles 69
Metalloids 83, 142
Metals 76
Methane 5, 6, 161
Methanogens 165
Metropolis 3
Metropolitan areas 3
Microbes 5, 135, 143
Microbial 35
Microfinance 186
Micronutrients 22
Microorganisms 5, 22
Micro-plastic contamination 35
Microplastic particles 33
Micro tunnels 34
Migrants 83
Migration 156
Mineralization 30
Mineralogy 24
Mining 10
Mobility 11
Mobilization 105
Monitoring 76, 84
Mulching films 34
Murky water 171
Mycotoxins 22

N

Nanoparticles 84
Nano plastics 39
Natural killer cells 51
Natural predators 157
Natural resource site (NRS) 95
Natural toxins 70
Negotiations 175
Neurological 50
Neurotoxicity 41
Nitrate 5, 150
Nitrification 165
Nitrogen fixation 130
Nitrous oxide 6, 7
Nitrous oxide emissions 33
Nobility 111
Non-biodegradable 71
Non-biodegradable materials 128
Nongovernmental organizations (NGOs) 170
Non-isotopic 162
Non-point pollution 130
Nontherapeutic 160
Non-toxic 139
Numerous spillages 121
Nutrients 96, 179

O

Obsolete pesticides 26
Oceans 89
Organic compounds 13
Osteoporosis 12
Owners 122
Oxy-degradable polymers 35

Oysters 19

P

Parkinson's disease 50
Pastures 96
Pathogens 5, 39, 92
Perfluoroalkyl compounds 39
Pesticide remnants 128
Pesticides 2, 23, 117
Petroleum hydrocarbons 134
Pharmaceutical drugs 22
Pharmaceutical products 82
Pharmaceuticals 32, 101
Philosophy 130
Phosphate 5
Phosphate rock (PR) 28
Phosphorus 13
Photo-degradation 36
Phytoextraction 136
Phytoplankton 63
Phytoremediation 135
Phytostabilization 11, 136
Phytotoxicity 73
Phytovolatilization 136
Planting 176, 179
Plastic mulch 34
Poisoning 2
Politicians 167
Pollutants 170
Polluted zone 141
Polycyclic aromatic hydrocarbons (PAH) 139
Polyethylene 35
Polymer 35
Population growth 22
Post-remediation rehabilitation 139
Potential remedies 182
Poultry 113, 159
Poultry products 72
Procedures 34, 135
Producers 158
Production 8
Production process 41
Productivity 106, 108
Profitability 108
Pseudo-cofactors 12
Public-Private Partnerships 112

Q

Quantity 128

R

Radioactive elements 82
Radioactive waste 6
Ranchers 95
Rapidly 141
Receptors 174
Recognizing 150
Recreation 14
Recreational facilities 90
Recycle 37
Reducing plant 101
Reforestation 129
Regulation 119
Regulators 158
Regulatory process 174
Remediation 174
Reproduce 62
Respiratory disorders 62
Restoration 15
Rhizodegradation 136
Rhizofiltration 11
Riparian zones 96
Rivers 89
Rubbish 170
Runoff 184

S

Safety violation 121
Salinity 130
Scientists 44
Sedimentation 96
Sediment, nutrients 2
Sediments 150
Seed germination 35
Septic sludge 32
Sewage 170
Sewage system 129
Sexual maturation 51
Silt disposal 96
Slash 177
Slaughterhouses 159
Smoke 157
Smokestacks 3
Snails 163
Soil alteration 133
Soil biopiles 143
Soil contamination 82
Soil detoxification 129
Soil erosion 97, 179
Soil fertility 170
Soil microbiome 185
Soil permeability 146
Soil quality 128
Soil transfer 95
Soil vapor extraction 141
Solubilization 104
Sorghum 185
Sorption-desorption processes 130
Specialization 157
Spills 93
Stability 108
Stakeholders 119
Steroids 33
Stormwater 68
Streams 3
Suburbanization 14
Sulfate reducers 165
Sulfur dioxide 3
Supplier 118
Sustainable energy 129
Swimming 92
Synthetic fertilizers 44
Synthetic industries 10
Synthetic inorganic 26

T

Tax liability 112
Temperature modulation 134
Thermal decomposition 13
thical conductE 171
Thinner shells 62
T lymphocytes 51
Topography 172
Toxicity 12, 174
Toxins 17, 123
Trace elements 40
Tracer methods 162
Transformation 8, 170
Transforming data 99
Tumors 17
Typhus 23

U

Unemployment 116
Uranium 82
Urban areas 3
Utilization 130

V

Vadose layer 141
Vadose zone 142
Vaporization 184
Vegetation 172
Velocity 95

Viruses 2
Volatile 142
Volatile hydrocarbons 135
Volatilization 163

W

Wastewater 71
Water hyacinth 92
Water percolates 183
Water pollution 2
Weathering 36
Wetlands 95

X

Xenoestrogenic compounds 78
X-ray fluorescence 84
Xylene 146

Y

Youngsters 182

Z

Zooplankton 63